Shorty and the Radio Men

A Young Writer's Light-Hearted
Encounter with World War II, Written
in 1946 and Not Released Until Now.
Authentic, Richly Detailed,
Definitely Not Sanitized

Myron Sutton

Illustrations by the author

Silver River Books
Medford • Boise • San Franc

This Silver River work contains
the complete text of the original book.

SHORTY AND THE RADIO MEN

First Edition May 2000

For information, address:
Silver River, Inc., 1619 Meadowview Drive, Medford, OR 97504, USA
The Silver River web site address is www.silverriver.com

Library of Congress Catalog Card Number 99-96712

ISBN 0-9662817-5-6

PRINTED IN THE UNITED STATES OF AMERICA

10 9 8 7 6 5 4 3 2 1

BOOKS BY ANN and MYRON SUTTON

Eastern Forests: An Audubon Society Nature Guide
The Audubon Society Book of Trees
Wildlife of the Forests
The Wild Shores of North America
The Pacific Crest Trail: Escape to the Wilderness
Wilderness Areas of North America
The Wild Places: A Photographic Celebration of Unspoiled America
Yellowstone: A Century of the Wilderness Idea
The Secret Places: Wonders of Scenic America
The Wilderness World of the Grand Canyon
New Worlds for Wildlife
The American West: A Natural History
The Appalachian Trail: Wilderness on the Doorstep
Among the Maya Ruins: The Story of John Lloyd Stephens and Frederick Catherwood
The Life of the Desert
Animals on the Move
Journey Into Ice: Sir John Franklin and the Northwest Passage
Guarding the Treasured Lands
Exploring with the Bartrams
Nature on the Rampage
Steller of the North

INTERNATIONAL EDITIONS

The Audubon Society Book of Trees (Japanese)
Le Monde des Arbes
Walder
Bomen uit de Gehele Wereld
Wildlife of the Forests (Japanese)
Knaurs Tierleben im Wald
Animaux des Forêts
Dierenleven in de Bossen
I Maya
Tiere Unterwegs
Hur Flyttar Djuren?
The Endless Quest
Irrfahrt im Beringmeer

The Barry Ross International Mysteries

The Isis Command
Death In the Amazon
The Madman of Mount Everest

(in preparation):

The Chala Project: Murder in the Grand Canyon
The Dinkum Deaths: Murder on the Great Barrier Reef
Murder in The Alhambra
Barry Ross and the Tiger Bone Gang
Murder at Angel Falls
Murder in the Galápagos
Death Among the Gauchos
Murder in Yosemite

Contents

"Was there much kick?"

2000
INTRODUCTION

I knew Shorty better than anyone. He was, at 19, an enthusiastic kid, eager to go places and learn, learn, learn. He kept records, wrote long letters home, and as soon as he got back from the war, unloaded his duffel bag full of maps, photos, clippings, notes, schedules, and brochures, and started writing—on a portable typewriter. He recorded faithfully every experience he could think of, in chronological order, about the radio men, of which he was one. Between 1946 and 1950 he produced a book of 221 typewritten pages, single-spaced, with one carbon copy, and titled it *Once Is Enough.*

And that was enough. He put each book away, in a safe place, and went on to write other books (30 at present count). Once in a while, he would get out that first book and reminisce about his adventures as a combat radio man, and when his own sons grew old enough to absorb the language, he would read parts of it to them. Those sessions didn't last long, because the sons would burst into hysterics and exit the room choking. He had taught them to have a sense of humor.

In going through the kid's typewritten first draft of more than fifty years ago and preparing it for public consumption in today's world, I have had to make a few changes. There are certain conventions regarding race and gender, for example, that have come to be abhorrent but were still in common use then. I must rob Shorty of his innocence on those points.

He made too much of the verb *to be*, and too little of the comma. I realize that these were minor matters, but this was the first draft of his first book, so we can allow him only so much latitude.

Also, every word of this book is true, to the best of my knowledge. Because of that, I must change the names of the principals. Supposing that Winston Morgan, whom Shorty, because of Morgan's excessive flatulence, called "Winnie the Peu," became a high level banking magnate, a pillar of social consciousness in the Boy Scouts, and generally a paragon of virtue. Shorty recognized him as a born leader, all right. Not only was Winnie the champion farter of Hut 9 but he organized intra- and inter-hut competitions in the latrine to see who could stand farthest back and still hit the urinal. What kind of a capital would *that* put on Morgan's pillar of public fame?

And what about that cold February night on the French Riviera, just before we moved to the Front? Dody Ratigan, too sleepy to crawl out of the tent and walk the length of the company street to the latrine, discharged the contents of his bladder into his steel helmet and placed the bowl outside the tent. Next morning, the liquid was frozen solid, and no matter how Dody tried, he could not dislodge the ice in time for morning muster, when every soldier in the war zone had to have on full field equipment, most importantly the helmet. What of Dody's reputation?

In mercy, therefore, I cannot let Shorty resurrect the names along with the events. I realize that this should be of no concern to us now. In those days, it mattered far less who you were than what you did. But that was then, and this is now.

Shorty's is an authentic, if somewhat romanticized, tale of where they went and what they did. And it is also a *comédie d'erreurs* about how a gangling teenager six feet five inches in height survived in an Army that, he swore, catered to those five feet nine. And then, from this comic tale, he wrote his first magnum opus when he was 21.

Yes, I knew Shorty very well indeed, his innermost thoughts, yearnings, frustrations, and victories. Why not? He was me at 19. And I can assure you that had he not written what he did, fifty years ago, this immensely detailed sample of history, Americana, nostalgia and war would have been lost and nearly all forgotten. Shorty gave us a window onto a turbulent time... just as the turbulence ended. He thought it important enough to write down, but not much more important than that. At the time, he wanted to forget about the war and get on with his life, college, career, and other writing. In retrospect, the era he described has been magnified if only because so much of the average GI's life of that day has vanished. He cherished those adventures.

He still does.

Myron Sutton

Obstacle course.

1946
INTRODUCTION

If I ever have any grandchildren, I'm afraid my knee will be too weak to support them. They'll have to shut off the television, curl up on the rug before the atomic fireplace, and read this which I have set down in lieu of hot air. For this is the kind of stuff you tell the kids, though in this case they'd better be well into their teens.

What is written here is not really important because I didn't have a whole lot to do with winning the war. There were millions of us in the thing, and those millions did far more than I to beat the enemy back. As a matter of fact, I was in high school when it all broke out, and I didn't have the slightest thing to do with it until early in 1944. That was late enough, you see, to be positive that the Allies were going to win the war, and it was, to me, just an experience.

But an interesting one, to my way of thinking, and sometimes very exciting, even dangerous. You'll find little blood and tears here, no prolonged suffering and profound agony, as was the lot of those who really deserve to be talked about. I just reflect upon my own career as a radio man and draw from it that which you are about to read. To say that I liked Army life is not saying the whole story.

Come with me, then, to the foothills of the Blue Ridge, to Georgia, the Ozarks, the waters of the wide and dangerous Atlantic, the shores of a strange land, the Riviera, the Rhine, and all the rest of it, and I think we'll have some fun. Slog through the mud of Morrisville and St. Jean Rohrbach, chat with Madame Joan, enjoy an ecstatic moment on the bank of the Lahn. Yes, I liked Army life... but...

Once is enough.

Myron Sutton

ONE
MACARTHUR

Today's Santa Fe trains out of Arizona into California carry travelers enjoying maximum comfort and ease. As the Scout, Chief and Capitan whisk across the desert country at subsonic speeds, faces peer out of windows to enjoy the bright red cliffs, vast desert wastes and mysterious dark mountains. Inside, there are comfortable lounges by day, luxurious berths at night. The thirsty stroll back for a cocktail, a martini in the lounge. They relax in an easy chair and watch the land go by. Children occasionally play up and down the aisle. A porter saunters leisurely through, securing comfort here, helping there. Riders revel in the good food and quiet atmosphere of the dining car; they even enjoy somewhat strolling back or forward to it. Friends chat. Bridge, pinochle, poker once in a while. Today, indeed, the trip is a pleasant one. Life is easy. "Getting there" is fun, relaxing, restful. It wasn't, however, always so.

The only comparable thing about the route when I made the trip was the landscape outside. The bright red cliffs, vast desert wastes and mysterious dark mountains still went by the window. The vestibule window. I think at first I enjoyed standing in the vestibule to watch the scenery outside. At length the novelty wore off. One reason for standing in the vestibule of the car was that there wasn't any place else to stand. Oh, I suppose I could have gone back into the car and stood in the aisle or washroom, but here in the vestibule I could stand up straight and that, at least, was something.

To put it mildly, the train was crowded. Every possible seat, of course, was filled—not only the regular seats but such choice places as the floor space, window shelves and wash basins. There had to be rotation in the rest rooms, though, or those seats would have been taken. For exercise, I walked through the train. In doing so, I occasionally ran into my cohorts who had shipped on with me at Flagstaff. They were out for exercise, too. One of them mentioned that about a half hour back he'd gotten to sit down for about five minutes. The plutocrat. At that, there were many besides us who were out for the walk so if we felt uncomfortable, we didn't feel alone. Rear ends stuck out into the aisles, seeming to bear the admonition "Don't Tread On Me," but that was expecting the nigh impossible. In walking down the aisles we had to kick, shove, plow and plunder. We stepped over, crawled under and came up gasping for air.

Occasionally, we found that some fool had brought coffee into the car and some other fool had upset it. This made the floor slippery. I remember sitting down once this way, but they made me get up so as to clear the passage.

It wasn't a troop train. We wished it were. There were servicemen on it, though. Soldiers, sailors, marines, WACS, WAVES… There were all manner of civilians, bawling babies, juvenile delinquents, a melting pot of the last 2,000 miles. Miles many people couldn't travel by car because of gas rationing. Most passengers sat twisted, crouched into the last inch of space, much like sardines, to coin a phrase. When night came the contortions of sleeping forms were something to see. Around midnight, we found ourselves in the dining car. Here many passengers draped over tables and chairs, sound asleep. They'd have to leave by breakfast but this temporary bedroom became a haven of peace. Miraculously, we found a couple places to sit but it only lasted half an hour. I went back to the vestibule.

We hit Barstow about dawn. By that time, the vestibule had become a dungeon, but it was still preferable for its fresh air. I sat on the hard steel floor and tried to sleep but gave that up because the floor was hard steel. All night long, the cracking of wheels on track and scrape and squeal of the car connections had kept up an incessant rhythm but they say you can get used to almost anything. By dawn, a thin film had slid down across my eyes and I didn't much notice the green valleys and canyons through which we rode. In fact, I didn't notice much of anything. Later, I found myself asleep in a chair in the chair car. How I got there and who I murdered for the seat I will never know, but at the time it was of little consequence. I rolled over in the seat and looked groggily out the window, saw spots and turned back to sleep. Then I realized we were in the orange groves that the Los Angeles Chamber of Commerce says cover all southern Cal. I tried desperately to prop up one eyelid to watch the scenery for a minute, but that was no use and the next thing I knew was LA.

The train rolled into Union Station and I became awake. I do remember crossing the Los Angeles river and asking someone where the water was, but I don't think I was fully awake until the train stopped. Apparently I'd slept enough to remove the thin, cloudy film because I could see where to step when dismounting. A wave of relief came over me at stepping off that train. I found the other boys, and we spent a couple hours in the station getting cleaned up, eating and looking around. We saw several groups of new troops fresh from Fort MacArthur on their way out to training camps somewhere. Young kids mostly, a few post-Pearl Harbor fathers who hadn't been able to stall any longer, all spic and span in fresh uniforms that were still creased where they'd been folded, and in brand new helmet liners. They lined up clumsily in front of their train and moved awkwardly when brought to command. They were rookies and we had to laugh. We weren't even rookies yet, and were having our last laugh.

It was the fourth of February, 1944, and there was little sign of winter in LA. To us, it was like summer in Flagstaff. That afternoon we found our way to the electric car terminal and climbed aboard a car for the 35-mile trip south to San Pedro. We tucked our suitcases underneath the seats and not long after that the train took off south. Going down along the coast of Southern California was much in contrast to Arizona traveling. It wasn't in and out of one town and then another; we just never left civilization. There were houses and fields and buildings all the way, with roads and cars and canals and a few trolley stations. Off in the distance were a number of oil derricks and I think we even saw Signal Hill, at least we saw a veritable derrick of forests, or was it forest of derricks—we may have dozed once or twice on the way. The most desirable thing about the car we rode was that it wasn't crowded.

We couldn't keep from thinking about MacArthur, and what we conjured up about our first days there was anything but delightful. We talked a while on it and swapped tales of what we'd heard about Reception Centers and the kind of reception they had to offer. We saw needles coming at us from out of the darkness, sergeant stripes, marching and camping, even combat, guns, shells, bombs, horrible deaths. *Our* horrible deaths. But peering into the future that far away wasn't what concerned us, for right then we began to live in the present—as the Army indoctrinates you into doing—and were only considering the morrow at MacArthur. From then on we signed a quit-claim deed to our bodies and souls. We breathed our last of freedom and knew it. I have a pang of pity for myself because at the time I was kind of excited and looking forward to it. Maybe it was fun. I don't know.

The Pacific Electric Railway terminal at San Pedro isn't much more than a waiting room at trackside. When we pulled in there wasn't any sign of the customary brass band or waving crowds. On a bench in front sat a couple of disgruntled sailors and around to the side was an Italian selling peanuts and a couple of girls the sailors were ignoring. San Pedro itself consisted of a half-busy street and tall, closely-crowded buildings. It wasn't clean but then there was a navy base nearby and an Army camp even nearer. Probably Long Beach had drained San Pedro's self-respect away. We could see Long Beach through the windows on the other side of the car. A narrow harbor contained a number of gray-painted ships, obviously brought in for repair for most of them looked well weatherbeaten and some were drydocked.

The car ground to a stop and we lugged our baggage off only to find that no one was waiting for us. We sauntered around long enough for the lead gray, cloudy sky to turn black, and for the street lights of San Pedro to begin blinking on. We began to think that the Army had forgotten us—at a time like this you grab at any straw—and even got to thinking of the next car back. But we decided that Federal Penitentiary, for all its safety, wouldn't be as wholesome as MacArthur, and sat down to wait.

By now, there was a little group of eight or ten of us waiting and just about through dusk an Army truck pulled into the station. We climbed aboard. Five seconds later, we wished we could climb right back out again.

Gas rationing in civilian life doesn't prepare you for the wild, devil-may-care, to-hell-with-the-gas type of driving we met that night. We got our first taste of GI driving.

It was only a mile to the Fort but it seemed at any moment to be the last mile. The driver apparently forgot all about us or thought we all had good, healthy heart conditions. He took one corner so fast I was pinned head down against the side of the truck and almost blacked out when we came out of the dive. The floor of the truck was muddy and on another corner one poor soul riding with us gave up the ghost and slid off his seat into the mud. But we got to MacArthur.

At this late hour we found offices vacant. We were herded into a vacant orderly room and there we

K 3315551 5 5
UNITED STATES OF AMERICA
OFFICE OF PRICE ADMINISTRATION
RATION COUPON
FOR
FIVE POUNDS
SUGAR
5 OPA Form R-325 (Rev. 2-44) 5

presented our credentials. Naturally they were in order, so we were assigned to Barracks 13, Company B. They gave us a comforter, two blankets, two sheets and a pillow case and took us over to Barracks 13. We made beds and though we'd liked to have dropped into them, we thought it better to drop with a full stomach so we searched out the mess hall. The cook was writing a letter, but after hearing our sob story he brought out some bread and a couple pounds of butter and some jam and other leftovers, and we ate. And sugar, without a ration coupon! It was too much.

We wandered around a little and watched MacArthur night life. The PXes were noisy, there was a mile-long line in front of the theater and millions of rookies roaming about, probably in a daze. All sorts of activities were going on in the barracks we passed—crap games, letter writing, shooting the bull, shoe shining, reading, eating ice cream, sitting and thinking of home.

Since they'd told us tomorrow would be a busy day, we meandered back to the barracks. After a heavenly shower we crawled into our bunks. That thin cloudy film was beginning to make everything hazy again, and so, amid the noise and confusion of the room, we dropped off. I believe I went to sleep just as I was crawling into bed. My shin didn't bother me—tomorrow I'd find another way to get into the top bunk.

The next day I had the time of my life. This was the Army, Mr. Jones! Up and out of the bunk bright and early at six a.m. and drop dead in the washroom. My civilian life had been going to bed at two a.m. after working late as a motion picture projectionist and then getting up, oh, around ten the next morning—whenever I felt like it.

This first morning in the Army, a loud whistle blew over the loudspeaker system and I like to fell out of bed. Not having yet learned the technique of dismounting from the aerial perch of an upper bunk, I strained my sleepy, flabby muscles and twisted into a contortionist act to jockey into position for the descent. The fellow in the bunk below, if indeed he were fool enough to be awake yet, must have gotten seasick. But finally I found a good take-off position and let go, crashing to the floor in what I thought was a heap of broken bones and mangled flesh.

Breakfast compensated. It was out of the civilian world, a pre-war dream come true again. All the solid, 14-carat butter we could stuff down, and milk by the quart. Army life did have its advantages.

After chow we cleaned up the barracks spic and span, with what little knowledge we had of housekeeping. We fell out then in the company street and the sergeant came to take over. He was a youngish fellow, obviously bitter about not being able to get over in combat and perhaps a little jealous of our getting the opportunity to go to our deaths. He wore sunglasses so we could never tell just whom he was sizing up for penalties. However, he did show wisdom in one respect: he made no attempt to have us march.

That first morning was devoted to preliminary details. We received the old GI gospel, the Army Service Field Manual, and some other materials that would require days to read. Then we had a slight physical, that is to say, they determined whether or not we were breathing.

The afternoon brought tests, the AGCT (Army General Classification Test) first, which types you and opens the gate to whatever branch for which you're best fitted. Months later I found out my score was 128. That was genius, I thought, until someone told me. The radio signals test brought more concentration out of me than anything else. I was shooting for Air Corps radio, but then, what you want and get are two different things they say.

The boys in the barracks were shooting craps. They said some guys would win $100 and send it home. On the other hand, some would lose $100 and write home for money. I could think of a hundred other ways to lose money more profitably, and during my Army career tried out a number of them.

The VACS canteen was a pleasant place to sit and guzzle free Cokes and root beer. It was a big place, with tables, couches, piano, and card tables and several hundred GIs. But though it was packed most of the time, letter writing wasn't difficult and occasionally the typewriter would be free. That first night, some southern boys sang in one corner with piano and bass fiddle, and they were good.

Late that evening I walked out away from the barracks and stood against the wire fence looking out on the lights of the bay. Heavy clouds increased the darkness to point up the glitter of the lights and shimmering reflections on the water surface. Some red and green lights added a Technicolor glow to the scene, and now and then I could hear the muffled, mournful groan of a buoy out in the bay. That piece of water out there, small and quiet as it was, seemed the open gateway to the trade markets, glittering beaches and savage battle grounds of the world. It suddenly made the world a very small globe indeed. It was the pathway to that which was unknown and yet the very serenity of the scene gave me to feel that that pathway was not or would not be one of bubble, bubble, toil and trouble. There was no fear or disheartenment here on the waters and lights of San Pedro Bay, and I liked it. Perhaps the thousands like me who had passed this way—those who had scarcely been away from home before—felt as I did: a pleasant apprehension, a compelling desire to get started on the road ahead, to learn new things, to see new places, to meet new friends, to take on new interests. And as I stood there, face pressed against the wire, I slowly reassured myself that this whole business was going to be one helluva lot of fun, from beginning to end. That may sound pretty dumb, but it was better than being morbid about the whole affair. No matter what guard duty, KP, inspections and bawlings out may come, ten years from then it would all be one big laugh. I guess I was incurably upbeat.

The next day was unforgettable. We weren't, of course, prepared for the first shots we got. One in each arm. We stood placidly, innocently waiting for the medics to do what little damage they might. Naturally, they wouldn't dare hurt us. That was against Army regulations. And then they hit… one just after the other, stabbing with needles as big as tommy guns and twice as deadly. After the initial shock I didn't feel anything. Anywhere. Then rigor

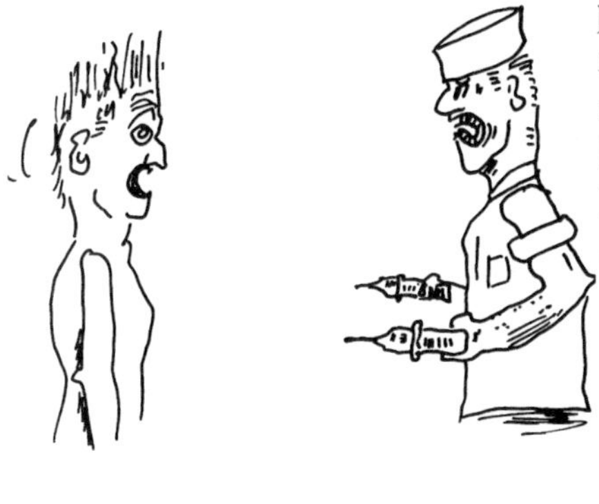

mortis set in, up one arm and then the other. One fellow just ahead of me turned white after the shots were finished and came falling backward. None of us was in a position to hold anything up, so he plunged to the floor. He was back with us in five seconds, though, and we all laughed about it. What was so ridiculously funny we didn't know, but that was the way we passed it off.

Then we received our first Government Issue:

1 razor, safety	1 blouse	1 kit, cook	1 bag, barracks
5 blades	1 raincoat	1 knife	1 bag, traveling
1 case, razor	1 towel, hand	1 fork	2 shirts, OD*
1 toothbrush	2 towels, face	1 spoon	2 pants, OD
1 comb	3 pants, suntan	5 pr shorts	1 belt
1 liner, helmet	1 blouse, fatigue	3 shirts, sun tan	2 pr white socks
5 undershirts	1 pants, fatigue	2 pair shoes	2 pr khaki socks
1 pr leggings	1 canteen	1 overcoat, OD	1 pr gloves
1 hat, sun tan	1 cup, canteen	1 jacket, field	2 ties
1 hat, OD	1 cap, fatigue	1 case, canteen	* *(Olive Drab)*

One of the most surprising things about the articles of clothing was that they actually fit. Of course, I had to try on some things ten or twelve times, but nothing sagged or pinched.

The manner of receiving all these materials was one of those Army miracles of assembly-line efficiency. We received the duffel bags at the head of a long counter behind which stood a row of clerks all the way down to the far end of the warehouse. Holding the duffel bag just below the edge of the counter, we began to move, shouting out sizes and shapes and having each item thrown into the bag. By hitting our bag accurately many of them hinted at having been basketball stars. At first we went along aimlessly, holding our bag up to the counter. Then, as it became more and more full, the bag began to drag and we were catching things and stowing them away.

Everything went along fine, as far as I was concerned, until I came to the shoe counter. I shouted out my size. The clerk went white and looked up. "What did you say?" he asked, in a blank voice. I gave it to him again. He turned and yelled up the line: "Oh, Fred! Come down here, will you?" Then I was requested to fall out of line and sit on a bench to the side.

After a while, Fred came over and measured my foot. Then he went away for a moment, got another measuring device and came back to check. After that, I began trying on shoes. First, too small, then too wide, then too long, then too narrow, until at last, after an exhaustive search of the camp shoe stock, I got one that was just right. Size 13. That was the first indication I had that the Army wasn't, er, ready for me.

Fred brought out some museum pieces to show. Size 15½. Fifty dollars worth of shoe, he said, probably required the better half of a cow hide. After that, they let me in gently on the bad news. Because I had such an unusually large size, I was allotted *two* pairs. Another 20 pounds added to my already overcrowded, bulging duffel bag. I couldn't believe it. I would go through the war with two pair of those shoes? Little did I realize then what else was coming to someone too big for the Army...

Next morning, several men were called up for the Air Corps. Not me. I began to realize that I probably couldn't have fit into a cockpit anyway. I resigned myself to waiting for the Signal Corps or some similar soft job. Maybe that would come later.

I had a pass for Saturday afternoon and Sunday, and wanted to go to Santa Monica, but I gave that up when someone gave me a list of directions on how to get there. Instead I YMCA'd in San Pedro and strolled along the park overlooking the bay. The vast expanse of the Pacific stretched to the south; eastward were the towering buildings of Long Beach and hundreds of ships. Beautiful, peaceful scene.

Monday I got my first GI hair cut and those fascinating red curls fell with tears all over the place. But I gave up and said to hell with it, I'll flash my handsome uniform on the ladies from now on.

Now we were soldiers of Uncle Sam's gallant fighting force, the best equipped soldier in the world, ready to burst forth in shining armament to meet the foe and destroy him in the rain of smoke and shells by the dawn's early light and the shining magnificence of Old Glory. Next day, I was on latrine duty.

The GI movies were a new experience for most of us. Especially the ones on sex hygiene. These brought on a lot of squirming during the parts that showed explicitly what happened *without* good hygiene. The movie entertainment of most GIs up to that time had not normally included detailed views of cankers, sores, and blisters in the wrong places, and so they left the theater considerably shaken.

Speaking of movies, while we were at Fort Mac we saw Donald O'Connor in a rookie line. One day we spent a whole afternoon as extras with Eddie Robinson and the Columbia film crew working on *Mr. Winkle Goes to War.* I thought I'd be discovered but their talent scout must have been off duty that day and none of the rest could recognize real talent. José Iturbi played for us one night and sure went over big. I think a little culture was good for us.

The first day on KP wasn't as bad as the books and movies would make out. But it was work and the individual GI's recognition of that term would spell the degree of distaste he had for KP. But this is not to say I liked it. We'd had some elementary drilling out in the sun on the asphalt parade ground and even that was much to be preferred to kitchen duty, but the inevitable had to come.

Arising at 5 a.m. was worse, naturally, than at six. Peeling potatoes I didn't mind, but mopping floors was the end. Most of the mops were made for short people and I got the affliction common among six foot fivers who mop kitchen floors: back pain. They called four of us big, husky guys out (when they're desperate they'll compliment you in any way to get the work done) and we worked a while with the vegetable truck. I'm not particularly cut out as a weight-lifter but the work wasn't hard and gave us an opportunity to play catch with lettuce heads. The compensating feature of KP is that the cook doesn't rap your knuckles for stealing a little extra dessert.

At last we finished processing, and with that began a period of waiting for the orders to come through. This opened up the pastime of speculation and many bull sessions were devoted to discussion of various camps throughout the country to which we might be sent. According to the general opinion, being sent to any of them would be like a lamb going to slaughter. Texas provided material for the worst comments and it was thought those who flunked the AGCT would go there. That little thought, at the time, had a number of us worried. And it didn't help matters any when they told us that if we moved around much, as was likely, we might not get paid for several months. I tucked away a three-cent stamp or two for that desperate air-mail letter home in such a catastrophe.

And so the waiting went. The Army food was beginning to tell on me. At this point I'd gained ten pounds since the induction physical. Of course, I'd be working that off in sweat and tears in a few weeks, but it was nice to know what good square meals could do—together with regularity and sleep.

At length, the orders came through and the waiting ended. First off, they only told us that we were leaving. That had a maddening effect, but it also brightened our spirits again so we set eagerly about packing our bags and readying everything for the trip. Next day, we were lined up in rows with all our baggage and assigned to groups departing to varying destinations. My name was called, and I went over to my specific group—Alabama bound.

TWO
TROOP TRAIN

Union Station in LA hadn't changed much, except that now I looked at it with different eyes. This time I was among the rookies lining up and marching awkwardly to the car. We all had our helmet liners on and neat, newly pressed storage khakis. Most of us struggled under the load of GI equipment we'd received. I looked across to some of the waiting benches and, sure enough, there were some civilians on their way to MacArthur, having their last laugh, at us.

But then, now that we were leaving it, Fort Mac wasn't such a bad place after all. I think some of us enjoyed our stay there and it is left to be said that there are worse places to shake hands with the military. At least it was in sunny California, and we didn't freeze. However, if we'd known what was coming in the sunny south we might have taken along a sackful of that golden sunlight and pleasant warmth.

The troop train we mounted there in LA wasn't a particularly small one. There were 150 of us going to Alabama and that would require four cars. In addition, there were 200 sailors aboard with some eastern destination, and that was five or six more cars. If there were any beyond that, we didn't see because we were too interested in watching Los Angeles disappear in the sunset.

Having come aboard as we did near night time, there was nothing to do but to make our quarters ready for the journey and then sit around and shoot the bull till lights out and then drop off in slumber. Our bunks were assigned, and I could see at a glance that mine would be too short. I was pretty glad to have drawn an upper, for I concluded that up there alone I could angle my knees in any direction and not bother anyone. Of course, I might shove them out in the aisle and give someone a black eye, but that was his worry. For my part, straightening out in the bunk was certain to be impossible. The Army wasn't ready for me.

The next order of business was to find the latrine. No sooner said, than done—a tiny room at the front end of the car. I opened the door, ducked, and went in. It was the smallest room I had ever entered. It had a window of clouded glass, a corner bench, and a small, rounded wash basin. I couldn't even have bent over to brush my teeth, much less—.

Come to think of it, where was…? I stopped, looked around, blinked my eyes, looked around again. Where was the toilet? The sign outside had clearly said "Latrine." Weren't latrines supposed to have places to sit and…? I looked around hurriedly, desperately. My God, there wasn't even a urinal! The process of merely entering the room had stimulated my bladder to start filling up in preparation for discharge. While I had a considerable reservoir, and was, in fact, rather proud of it, my system was not designed to hold back forever. And now I found myself in a latrine with no visible receptacle in which to discharge *anything*.

Except, of course, the little dinky wash basin. I stood there, momentarily stunned. I couldn't believe my eyes. Something had seriously gone amiss. Was I *that* new to the Army? Must I look for a Latrine Number One in which to wash my hands, a Latrine Number Two in which to do Number One, and a Latrine Number Three in which to do Number Two? I stood there helpless and confused. How in hell were these troop trains designed anyway?

As urgency rose, I looked again at the wash basin... I had to set up a contingency plan pretty soon, because I felt that I had, at most, three minutes left. It was either the wash basin or the floor. The decision seemed to become clear. But, no, I couldn't do that to my fellow passengers. I could run out and try to find the shower car, but that solution risked leaving a trail of liquid down the aisles. How about the vestibule? Did troop trains have any? My time for research was dwindling.

The first stages of panic began to well up inside my fevered brain. I looked at the window. It had handles! I could lift it and hurl my stream outside. But we were going along at a good clip, and I knew what would happen. The spray would come right back into the room, and if it did, I'd have to seek more than a shower car.

There wasn't even a wastebasket for emergency use. I thought of flinging open the door and asking the first passer-by in the aisle "Where in the hell do you piss in this goddammed latrine?" But that would have been so mortifying, I couldn't bear to think of it. The kind of thing you get nicknamed for. No, this was a private matter between the latrine and me, and I wasn't going to bring anyone else into it.

Having now about a minute to go, I began to examine the bench. Why would anyone put a bench in a latrine? It had a handsome oak top, almost like an old-fashioned table top. If I had been able to bend over, I would have looked in the corner for some kind of latch. Perhaps it moved, revealing a hole down to the tracks. That would have been ingenious. So I pursued that possibility further by grasping the seat and seeing if by any chance it would move. At that point I discovered that the seat was, in fact, a lid, and when I lifted it, there was the familiar enclosed basin, and even a button for flushing.

Two minutes later, I withdrew nonchalantly from the latrine, and sauntered amiably down the aisle, rested, relieved, and flushed, you might say, with victory.

Our talk that evening turned to Alabama. No one around seemed to have ever lived there, so we were at liberty to conjure up what the country must look like and how it was going to be to have basic training there. Crawling among the copperheads. Itching with poison ivy. Hosting a horde of ticks. We were headed for the Infantry Replacement Training Center at Fort McClellan, 60 miles north of Birmingham. The thought that being trained as infantry replacements didn't strike us pleasantly. We summed it up in neat little phrases. Some guy goes over to the battlefield and gets shot. The Allies need an infantry replacement to take his place. And get shot.

We didn't ponder much on this, though. The talk soon turned to thoughts of Alabama women, those southern belles of Old Dixie. That was a far more pleasant notion of the future to take to sleep with us. Thus, with lights out, and our beds made, we crawled in.

This last was far easier to say than, at the time, to do. By the time I'd left MacArthur I'd managed to gain a satisfactory degree of efficiency at crawling into the upper bunk. Feeling that the method applied there would work as easily here, I at once swung up toward my berth, and promptly fell to the floor.

It took a combination acrobat, trapeze artist and judo expert to gain a half-Nelson on the upper bars and hoist myself up, Tarzan like, into the bunk. But when I finally got up there, I found my feet and legs in the bunk, my body swaying dizzily out over the aisle, and my hands firmly clenched on the support rod on the opposite side of the car. Not having been in this position before, I lacked the ingenuity of having escape methods at my fingertips. I just hung there. I'm sure by this time everyone else was sensibly asleep, so I didn't dare yell for help. Nor could I hang there and wait for the next porter. One might not come by before dawn. That left two courses open. Either I must let go and write my epitaph in a puddle of grease on the floor below. Or inch myself into such a position that I might swing back and grab the support rod on my own property.

I selected the better of the two choices and finally worked my hips into my berth. I dared then to let go with one support hand, grabbed quickly home with it and pulled myself in. This little exercise before sleep made me set about earnestly in disrobing for bed and I discovered then that I had made a serious mistake. Any normal fool would have undressed downstairs, thrown his clothes up and crawled in after them—a process I practiced the next night and on all troop trains thereafter. But on this first upper berth night, as I began to take off my shirt I rammed my fist into the wall with a jolt that must have been heard throughout the car. The subsequent squeal I gave must also have been heard, but no one complained. I gathered my wits enough to subdue my own stream of cusswords that followed this mishap. I got one arm out of a sleeve, then put both arms behind me to get the other arm out and the shirt off. In doing so, I pushed slightly on the wall and went sailing out into the aisle; but fortunately one free hand managed to grasp the support rod opposite, and I found myself in the same position as before. But this time, I knew the process and slowly climbed back in.

With the shirt free, pants came next. After immediately stubbing a toe, I used more caution with the remainder and was soon ready for sleep. So I lay down.

I don't believe that I have since whacked my head harder than I did that night, but eventually I scooted down, rammed my legs up double and was at last in bed. What a feeling! Shortly after that I knew nothing until morning.

When I swung down eagerly in the morning, I almost stepped on somebody's head, but that was a common thing on troop trains, and later I was kicked in the face while walking down the aisle… so nobody thought much of it. I decided to dress in the aisle now and make it much easier on myself. That would have been all right, but five other guys were trying to dress in the same place at the same time.

Performing one's toilette when everything around is jolting in all directions is not a task to which one may accustom oneself quickly. But having in some way accomplished this feat, I went down the aisle and stood in line for the wash basin. Performing one's ablutions aboard a fast moving train is an equally difficult task and I think sheer miracle prevented a face full of shaving cuts. There were several times I almost rammed my face into the mirror. Someone may someday make a fortune on padded railway tracks.

Being a troop train as we were, we made very few stops, and that was a good thing. Especially when breakfast came. We filed into the kitchen car and picked up our plate of food and staggered back to the car. A lot of good food was lost this way. But the food was good. That first breakfast consisted of orange juice, cereal, milk, toast, bun, two eggs, potatoes and apple sauce.

Beyond the little incongruities of the trip, it was an enjoyable ride. We waked in Arizona and after breakfast sat back with windows open to absorb that fresh desert air and watch the cactus float by. We arrived in Phoenix at ten o'clock.

Leaving Arizona again was no fun, but I stayed glued to the window, through New Mexico and into Texas. The Pecos and Devil Rivers were beautiful and some side canyons provided breaks in the scenery. On the train itself, some of the troops dozed, some read, some played cards, others talked. We passed through Uvalde, Vice President Garner's home town, and a couple of beautiful girls came up to the car windows and gave out little sacks containing gum and home-made cookies, home-made chocolate cookies filled with nuts. What a treat! We accepted with heartfelt thanks. Some of the passengers wanted the girls as well as the cookies, but we had to move on.

That night we went through San Antonio. Next day, Louisiana. By this time, most of us had become afflicted with a new scourge: homesickness. At practically every stop we saw men jump off and grab their loved ones and greet their friends. To counter our envy of these lucky ones, we spent time writing letters home, ending each with some version of "Having a wonderful time, wish I were home."

The merciless swamps and bayous of glorious Louisiana zipped past and we approached New Orleans. For hours, we rode beside sloughs, small patches of cypress trees, and extensive rice fields. We thought about amphibious warfare. The train stopped in the open once and we got out to stretch our legs. We found the entire track bed composed of small sea shells. The ride continued, and soon we crossed the Mississippi River. Not many of us envied the residents of the little shacks built in swamp lands below the level of the river. We got two hours off in New Orleans and ate in a little cultured greasy spoon [a second-class diner] just off the main thoroughfare. After strolling around a bit on Canal Street, we headed back to the train and that evening rolled on through Mississippi, on the Chattanooga and New Orleans Railway. Next day, we entered Alabama.

Fort McClellan, Alabama 1944

THREE
McCLELLAN

We arrived in Ft. McClellan on the 15th of February, 1944, and an hour after arriving I was on KP. Sometimes events occur faster than we plan for. Actually, my being on KP was a mistake. I volunteered for it. I was still in that stage of my military career where warnings about volunteering for things hadn't yet sunk in. The few of us who did volunteer were told that if we took it now we would, of course, have it easy later. But then, whoever told us that knew we were rookies and would swallow anything. And we did.

Having KP right away merely served to postpone the confusion of getting located and settled. Then, too, it allowed our buddies to select the best beds in the barracks to which we would be assigned. This was Company A, 9th Battalion, 3rd Regiment, and was only a temporary assignment. From here, we were informed, we'd be assigned to the other sections of the training center for which we were best fitted.

The food we had those first days in McClellan wasn't quite as good as that at MacArthur, and we soon found that this was because we were getting deeper into the routine, that the farther you get from the outskirts of Army life, frequently, the worse conditions are. But I personally found little to complain about for the food was yet good and plentiful, and we knew that dainty and fanciful preparation could be left for civilian life.

The barracks here in Ft. McClellan were not strictly the general Army type. Being smaller than standard barracks, they were termed "huts" and that indeed was apt. Each was a one-story affair, housing 10 to 15 persons, their equipment and a stove. They had tarpaper exteriors and a stove pipe, and, built on concrete posts, were set two feet off the ground. Inside, each soldier was allotted about 32 square feet, which I found reasonably comfortable and convenient. We could almost reach everything we owned from one spot.

We received a shock during our first two Alabama days. What we had been expecting as the sunny South turned out to be a cold, damp, drizzling, typical Alabama February day. Since we'd all caught colds of varying degree on the train, this weather we now found ourselves in far from helped matters. I feared that if we'd all got to coughing at once it would have rocked the hut off its concrete foundations.

We initial KPs finished duty the evening of the first day and made our way to the hut that was pointed out to us. We didn't do much that evening but throw our gear in an empty cold and dark corner and flop into our bunk beds. Since no one else was in the hut, the rest of them having gone out to guzzle beer in the PX or to a movie or service club, the thought of sleep soon wore off and some of the kitchen ache went out of our bones. So we got up and went to the movie.

Next morning we were taken on a brief inspection tour to see what the Fort was like. An orientation maneuver like this helped us to get acquainted with the place that would be our home for the next five months or so. While we were training to be infantry replacements.

Ft. McClellan was a big place. This much we could see from our brief excursion. It was one of the largest military reservations in the Fourth Area, comprising 19,600 acres. Everything everywhere seemed to be on a big scale. The Post Exchanges and Service Clubs were immense. The huge Fort McClellan Amphitheater, they told us, was an ideal place to present entertainment and instruction to trainees on a large scale. There seemed to be a million huts and buildings and twice as many men. There were four theaters, swimming pools, bowling alley, officers clubs, some stores and shops and ample parade ground. Each area had its chapel. The main Post Chapel, known as the Silver Chapel, was a masterpiece of simple stone architecture. The main PX, boasted as one of the finest in the country, was a veritable department store. There was a large post hospital, a two story affair with beautiful landscaping. The Headquarters building was an imposing red tile-roofed structure and we understood that there was an induction station here, too.

Surprisingly enough, many of the lawns were green. Whether they stayed green all winter or not we didn't know, nor did we care to stay over to find out, but we didn't think so. Yet this was February. We were *already* in the middle of winter. It was another cloudy day, of course, but we could see the heavily forested hills in the distance. We established our position by reading a picture postcard on Anniston—adjacent to whose city limits Ft. McClellan was located. "Anniston, The Pearl of the South, nestling amid the foothills of the Appalachian Range of the Blue Ridge Mountains, is the county seat of Calhoun County. Anniston was founded by Samuel Noble, an Englishman by birth."

Most of the trees in the surrounding forest were hardwoods, but there were a few pines and this was a comforting reminiscence of Arizona. We noticed the lack of birds here. The only way we could account for this was to think of the amount of firing that must go on every day on ranges back in the woods. Probably no self-respecting bird had flown this way since the war began.

Such was Fort McClellan and environs, and I became convinced that I was gong to like it here. Naturally, I changed my mind from time to time but for the most part it was nice country. I'm afraid I could never say the same for the climate.

During a long lull period, we had a couple hours drilling. Now it is one thing to take green rookies out on the field and begin to train them in the rudiments of marching—that in itself has been exhausted in books and films as a source of humor and horror based on the ridicule of man's soul. But it is still quite another thing to take these rookies out on a half-muddy field. Doing an about-face was just a polite way of sliding to first base. Then we stood for five minutes and practiced right- and left-face movements, then climbed out of the holes we'd dug and did some flanking movements. At first, when the column went left, I went right and the trouble was that since you must look straight ahead, I didn't know when I'd left the multitude. The sergeant's shrill explosions brought me to a screeching halt and forced a dash back into place. One fellow next to me seemed never to be sure of his right and left directions. Once when we were marching side by side, the command was given to turn right. I turned right. He turned left.

CRASH!

That afternoon, we were taken to an office building for interviews. We were told that here at McClellan there were specialized training outfits and that our little group, by virtue of the high score we'd made in the AGCT, might be permitted to select some such field for our training. There were cooks, bakers, artillery and engineers, communications, clerks, chauffeurs, mechanics, and others on an imposing list. We would be allowed to make a selection, and if we passed the interviews successfully, if our record agreed suitably, if we were built right, and other obstacles were surmounted, then there was a 50-50 chance of getting into specialized training.

So I stuck to radio, and went in to my interview. The lieutenant asked if I knew what Army radio really meant. I said, "No, sir." And he said, "It might mean some time picking up a 60-pound radio and all your own personal belongings and running to a new position under heavy fire. Could you do that?"

I replied in a suddenly weakened voice that I'd still give anything to get in radio, and that with proper training I certainly could do as he said, or at least would do my best to do so, whatever the circumstances. Apparently this made an impression, and I was given to understand that I would probably get into the communications school.

Now this was a wonderful thought. As I left the building, I thought happily about going to school and learning a trade, launching myself on a new career. And now that I was so valuable, with my outstanding IQ and all, certainly they wouldn't waste time on putting a man of my talents through basic training. That would be ridiculous. To start school would be the only logical way to get the war over, so let some of the other poor unfortunate infantry men go through all that silly basic stuff. A week later I was having bayonet practice.

The day after the interview, I learned that I'd made the grade. I was assigned to the Fifth Regiment, a specialists training regiment located on the far western outskirts of the camp. This outfit I learned was composed of companies devoted to trades like cooks, buglers, pioneers (trailblazers), armorer-artificers, clerks, wire, radio and message center, intelligence and motor mechanics. But that didn't mean a thing for the first two months. Those first eight weeks would be a stiff, concentrated basic training course that would presumably make us combat serviceable. So I resigned myself, as usual, fishing for the positive. The exercise would do me good.

So, on the evening of the third day in Ft. McClellan, a small group of us transferred from Co. A, 9th Bn. to Co. C, 15th Bn. And I don't think I'll ever know why we selected night time to make the transfer. We had to walk, of course, and night in a strange place is highly inducive to getting lost. But that wasn't enough. What made matters worse was that just as we'd gotten a block from our former home, staggering under the load of all our equipment, the worst rainstorm of the season commenced.

Few know what pouring rain, by the buckets that is, is until they've experienced an Alabama torrent. We were soaked to the skin in the first minute and then got lost. We somehow lost the pavement and plunged into an open field. An unpaved open field. Thus were we introduced to some typical Alabama mud, and as we sloshed along, a horrible thought struck me. I might soon be camping in this stuff.

We stumbled back toward some lights, dropped our packs, went into a hut to ask questions, came back out, hoisted our packs and took off again. Water dripped into our eyes, ears, collars and shoes. But by some miracle we at last stumbled into the

Fifth Regiment area and finally located C-15. The CQ issued our blankets and other material for the night and showed us to the radio shack. We unloaded our dripping clothes, dried off some and made our way across the street to shower. The dry sack that night felt good, with the rain still pouring outside.

Next day, we took care of the few details of getting settled in the company. It was a nice sunny morning, the first we'd seen and we were promptly put on detail, digging ditches and piling rocks. That evening the Radio platoon sergeant came in and chatted a while, giving us an idea of what to expect. Our cycle of training would begin in two days. First, we'd have drill, physical training, obstacle course, chemical warfare, military courtesy, Articles of War, and other little tidbits of Army life. The first week we would be in line for field packs, tent pitching, and bivouac. When he mentioned camping, I thought surely they would consult the weather office for a suitable day for this type of activity. Later I read the command sheet:

> INCLEMENT WEATHER SCHEDULE.
> All instruction will be held as scheduled.

That, it seemed, settled that.

Then, the sergeant said, we would combine our basic training in a week-long field maneuver at the end of the training, and, if we survived that, would then begin radio training. We eventually found out what he meant by "if we survived that."

I saw a group of advanced radio men with walkie-talkies working field nets, i.e., multiple radio transmitters on a single frequency. Watching their soft life turned my face green. But then, it wouldn't be long, I kept saying to myself. We checked out our steel helmets, cartridge belts, tent stakes and shelter halves and other field equipment, then sat back to wait for basic training to begin.

"WHICH left, sarge?"

FOUR
BEGINNER'S BASIC

We didn't wait long. Our cycle began on the 21st of February. That day saw military discipline, light combat pack rolling, preliminary rifle drill, radio school, and march, march, march. By the time we got back to the hut, most of us were in the crawling stage. We marched across the parade grounds to the theatre during the afternoon and there joined several other companies for training films. Apparently over 90% of those in attendance were feeling the effects of the damp, chill Alabama weather. Everyone seemed to be coughing with considerable vehemence. Colds were the common thing here. Some in our group had begun their training in a previous cycle, but had been taken out because of influenza and pneumonia and were now commencing again with us. Whether it would last this time or not they weren't in a position to say. Certainly half those in the theatre that day appeared to be in an advanced state of debilitation. As for myself, there was little cough, but the usual malady had inflicted itself upon me. I began to think of myself as "Sniffles Sut."

From the Army's point of view, the fact that training areas are located at a distance from quarters is a good thing. For us, in the early days, it wasn't that way. After breakfast we marched from two to ten miles to the point at which the day's work would begin. Good training, I suppose, builds leg muscles and the like, but a little pity on the poor souls we were at the time would have made us love the Service. However, I can't say I griped about it. High school days for me before entry into the Service had been filled with hiking and camping trips to remote mountain peaks and mile deep canyons. So, after the first day or so, the marching end of beginning basic became a snap.

Not so the calisthenics. Whether rain or shine, this process was accomplished without shirts. From time to time that was a pleasant way to exercise, but those times were few; the skies commonly were cloudy and frequently it was raining and cold. However, at the end of each calisthenics period, we were all steaming, and few of us complained about the cold.

The platoon sergeant who took us over and became our guiding light for the next 17 weeks was a man by the name of Stone, whom we promptly nicknamed Stonewall for reasons that will become obvious. He looked to be in his middle thirties. He was of medium size and build, with eyes none of us ever tried to stare down. I don't think I ever found out where he was from, though his accent could have placed his tap roots somewhere in Oklahoma. He was a good man—firm, harsh when we needed it, but over all he knew his business and was a pretty regular guy. He knew an endless number of exercises and kept pulling them on us. I used muscles I never even dreamed existed and they all got sore. Stonewall's cadence counting and our coughing made a sound that must have been interesting to hear. It might have sounded better if we'd coughed in cadence. Add to it the clank-clank of dog tags, worn on chains around our necks, every time we bounced. These little metal tags flipped up into our mouths and eyes, and banged us on the nose—but we got used to them.

When we marched from place to place, we would now and then see a company of advanced riflemen. They were pretty snappy as they glided along in smooth-flowing cadence, but there were invariably some stragglers behind who were the exact opposite. They limped along, dragging one leg after the other, looking like lost derelicts from the battle ground. There was distinctly nothing humorous about their plight, but we had to laugh. Of course, we could see ourselves in the same position later, but we laughed when we could those days.

There was considerable drilling at first. We polished off movements and commands and then got our rifles. The main idea behind this was that when we got to marching uniformly in flanking maneuvers, there would be fewer lost rifles. Just the same, there were those poor souls who forgot occasionally. When a flank movement was given, a sudden terrific crash would occur and everybody would duck while two rifles sailed through the air and plunged into the dirt. When two men with rifles thus collided, the impact was no small thing. I know. I was narrowly missed several times. Finally, though, as training progressed, we smoothed out considerably and these mishaps became rare. That was good. When we began drilling with fixed bayonets, not many ears were cut off.

"Don't worry, sarge. Just a few more weeks and we'll learn."

Stonewall had a stock statement, and we never heard the last of it: "You men have a high IQ or you wouldn't be in this regiment! Now use those brains and get these movements!" We got the movements. But there were some men who could never, it seemed, cure their club feet. I learned fast, and was often called out to help some of the less fortunate get in stride. For this advancement, I was sure, I would be rewarded in no time with sergeant stripes.

After every fifty minutes of drill or workout, we'd have a ten-minute break. During this period we relaxed, talked, smoked—sometimes we did pushups for infractions committed during the previous hour. At first, I made the mistake of lying down during such intermissions. But after breaking several bones, or so it seemed, getting back up again, I decided that the lesser of two evils was to stand in a drooped position, and rest at dinner time.

One of the first things we did after receiving our rifles was to go out on a four mile hike. During the first mile, the rifle gained considerable weight, increasing rapidly from the original nine pounds. My shoulder bone seemed to dislocate a little more with each step. First, just the shoulder became numb, then the whole arm, and finally the entire side was paralyzed. We would be given orders to change to the other shoulder eventually, thank heaven. By the time the hike was finished, both sides were in paralysis.

The first rifle practice was interesting. The governing principle was snap and precision: slap that rifle up, bounce it to your shoulder, get the hand away—fast. When the rifle was on the shoulder, and the order came to bring it down, we slapped the butt, twisted it around, and brought it to the ground. Some, however, slapped the butt and sent the rifle sailing ten yards off into a ditch. On the other hand, when the command "Right shoulder, arms!"—to bring the rifle to the shoulder—was given, some fellows would occasionally miss and sent the rifle flying back over their heads, hitting someone full in the stomach, or bounce the rifle off a helmet liner with the sound of a hollow bell. This encounter would make the recipient dizzy for some time, and cause the Sergeant to tear his hair and give us up as lost sheep.

One of the more difficult operations to perfect was that of inspection arms. This entailed bringing the rifle to the port position and snapping open the cartridge chamber so that an inspecting officer might look within for signs of dirt. The tensile strength of the spring retaining the cover of this chamber was such that if the thumb were not immediately withdrawn, you at once became four-fingered.

The most difficult part of rifle training, to me, was cleaning the damned thing. The bore, the sights, every little ridge, hole, and connection had to be made spotless, with rods, wire brushes, lubricant, and small cloth pads. Long after the others had finished, I would still be struggling to get out the last speck of dirt. By the time I finished, dirty rifle patches formed a circle around my feet. To complicate matters, the stormy weather made rifles rust continuously, and the whole cleaning process had to be repeated daily. When inspections came and went, I often found myself looking forward to additional duty. "Gigged," they called it.

And the rain continued. Mostly it was morning rain, with a steady downpour before noon. Surprisingly, the forest around McClellan didn't contain any appreciable swampland. Rivers running around and through the Post obviously took most of the flood away. And then, too, we absorbed a lot of the water ourselves, and dripped it all in our hut when we came home for dinner. Consequently, the huts were dank and humid much of the time.

One day we were subjected to a choice bit of torture that involved two shots, a vaccination and a blood sample. Immediately after that, there was a three mile march, half of which was double time (i.e., running in cadence). I think the idea was to make up some more blood, but it didn't work. Near the end, some of the less hardy fell out in the strict sense of the word, and appeared to be going into their last throes. Of course, by supper time everyone had revived.

Chemical warfare was rather a delicate element of our training. Drill and extracurricular practice made us proficient in donning the gas mask. Speed was the rule: whip out the mask, slap it on, squeeze, blow, grunt, and turn purple, just so long as it was fast enough to save our lives. When we were considered sufficiently prepared, we were sent through the gas chamber.

The first step in this exercise took place in an anteroom of tear gas—lacrimator irritant to them—so as to test the effectiveness of the mask. Then, outside again, we put away the masks. On signal, we lined up in front of the chlorine chamber, crouched like Olympiad runners, and shot in holding our breath. Once inside, all practice was virtually forgotten. Here, with horrible death all about us, we could do nothing but fumble. I remember getting my mask on at a time when I thought I'd almost surely explode from within. But our whole platoon made it through without a casualty and we retired to the open air.

"Help!"

Others, it seemed, had been less fortunate. They groveled under the pines not far away—racked with coughs that shook their inner souls—and ours. They were red in the face, crying and drooling, and the only decent thing to do was to pity them. But we couldn't keep from laughing at the contortions they were going through. We were ashamed, of course, because we knew that none of it was funny to them.

Later, several types of gases were released in the air so that we could identify them with a quick whiff. This finished chemical warfare training and I must say in all candor that thereafter we used our gas masks as filing cabinets for apples, oranges, letters, chewing gum, and what other bric-a-brac we might like to have at hand.

The inclement weather schedule was followed closely throughout. "Inclement" applied not only to rain and fog, but also to cold. The temperature may not have been down to any great low reading, but it was a damp, piercing cold that seemed to soak down to the marrow of the bones and stay there, no matter what measures were taken to warm up. One morning in the chow line, I started to freeze and that condition continued—prompted along somewhat by morning calisthenics. At a drill field adjacent to the battalion barracks, we removed our jackets, helmets and gloves—with an icy wind whipping about in sharp, jabbing gusts. I was glad when we started running around the field, for surely the exercise would have a warming effect. It didn't.

After an hour of exercises we were still as cold as ever. By the time we returned to our jackets, none of us could button them up. Our fingers had become numb. We got to go back to the huts for five minutes, back to a nice warm fire in the potbellied stove. Even then, the fingers remained too stiff for proper manipulation.

Then came an hour of rifle practice, with gloves on. That didn't help much. Army gloves are not known for maintaining warmth any great length of time. Then, happily, we had some indoor map reading and aircraft identification drill, but, of course, the building was cold. We were positive by then that Eisenhower would open a new front at the North Pole and we would be the shock troops.

But there were sunny days, too, and we reveled in them—basking and soaking up the solar rays every chance we got. On one of these calm, warm days we had a session in dirty fighting. Some of the principles of judo were applied, but most of it was just plain sneak, break, and kill. We were given to know that our current enemies did not follow the sportsmanlike rules of warfare and that we had to fight fire with fire. So we learned dirty fighting. This involved a number of sneak attack methods and painful holds, with nothing barred. One object was to kill or maim in the most effective, quickest and quietest way—just so we got the enemy first. The cadre performed several feats as demonstration and then invited us to join in the fun. I was enthusiastic. Here was a method of self-preservation that could be used anywhere, under any circumstances, and would come in decidedly handy sometimes. So I rushed in eagerly, to show the superiority of my extended height—and soon found myself flat on my ass. After being thrown a couple of times, I became much less energetic.

I decided to rest my fortunes on the trusty M-1 rifle and, wherever possible, avoid hand-to-hand combat. This decision was maintained only until I'd had a taste of bayonet drill. After that, I gave serious thought to making an application for some desk job 3,000 miles behind the lines.

The Handbook glamorized the science of bayoneting one's adversary:

> *The will to meet and destroy the enemy in hand-to-hand combat is the spirit of the bayonet. It springs from the fighter's confidence, courage, and grim determination, and is the result of vigorous training. Through training, the fighting instinct of the individual soldier is developed to the highest point. The will to use the bayonet first appears in the trainee when he begins to handle it with facility, and increases as his confidence grows. The full development of his physical prowess and complete confidence in his weapon culminates in the final expression of the spirit of the bayonet—fierce and relentless destruction of the enemy. For the enemy, demoralizing fear of the bayonet is added to the destructive power of every bomb, shell, bullet and grenade which supports and precedes the bayonet attack.*

Since I had not been raised to be a fierce and relentless killer, the Army clearly had a challenge in making me one. The object for which we strived in bayonet drill was fourfold: proper form, quickness, footwork and accuracy. Taking each of the four in order, I stacked up as a sad sack. No one with my build could maintain a proper form. Quickness was out, for to me the rifle always gained weight when we began bayonet practice. Footwork wasn't so bad, but then my shoe size was thirteen. I once took rhumba lessons and those had conditioned me for this process; but bayonet and rhumba steps do not bear a great deal of similarity. As for accuracy, I'm sure the sergeant would have liked me to go practice hitting a barn for a while, then come back to see if there was any hope.

The thought that my life might some day depend upon this knowledge led me to overcome some of the early weaknesses and I eventually managed to parry and thrust as good as the rest. A few men were slower learning even than I, and found their club feet back with them, but they, too, got the knack of it by the time we finished. Perhaps the most savage part of bayonet drill was the growl. Be aggressive, we were told—and so we learned how to hiss and growl with each thrust. Can you imagine thirty full-grown men jumping forward with rifles and grunting at the same time? The theory behind this was confusion of the enemy at close quarters. It confused us, too. But then, that was something to be managed in detail in the combat zone.

Our platoon was divided into three squads. The student platoon leader and each squad leader were selected from those among us who had had previous Army or ROTC (reserve officer) training. When these were put in place, each squad was then lined up according to height, with the tallest in front. This placed me next to the squad leader.

Right after that, a twist of events occurred in that the student platoon leader became ill and was removed from the cycle. My squad leader was selected to fill his place. This shoved me up into the front position. Thus I became squad leader quite by default. It didn't take long to act efficient and pull the wool over the sergeant's eyes so that I could remain squad leader and thus share the advantages that position offered. If any. It had drawbacks, however, such as the humiliation of inspecting my squad's equipment prior to officers' inspection and then getting gigged myself. But all in all, it was a good job and I became convinced that a responsible position like this was sure to earn stripes, one after another.

At times, the rain periods were so tenacious we called the Alabama dew a number of things of which the local Chamber of Commerce could not have approved. Mostly though, we called it Oregon mist: missed Oregon and hit Alabama. And each time we wished it had stayed in Oregon. But at last, on one of our field marches, someone noticed a flower growing by the roadside and we began to think of the coming of spring. That, indeed, was a lovely thought.

"Rest of my squad? Right behind me. Why?"

During most of the first basic training days, some time would be spent in classroom instruction periods, and in observing training films. There were classes in aircraft and mechanized vehicles recognition, malaria control, concealment and camouflage, military intelligence, map reading and the like. For obvious reasons, the first movies shown to us were those on hygiene, to put it politely. Then came the series of "Why We Fight," including the Battle of Britain, Battle of Russia, Battle of China, and when these were out of the way, the technical films were shown: camouflage, booby traps, combat carelessness, troop movements and so on through the list. Movie time during training was popular. It was a place of relaxation for escapists who wanted to get away from the regular routine. For others it was a good place to catnap. After each movie we were given a little test card on what we'd seen, but I doubt if these quizzes ever meant much.

We learned about Army organization, and the TOE (Table of Equipment) of the regimental units. This would have been a semester course in any normal college, but we learned it in a week. Or, more correctly, they tried to teach it to us in a week.

We had an introduction to the work of a communications company. One of our instructors was a lieutenant whose accent placed an "r" on words that had vowel endings. It was therefore a little difficult for us to keep a straight face when he began discussing the Chiner-Burmer-Indier Theatah of Operations.

Work in the field continued. Came the obstacle course. Now this bit of Army-contrived deviltry had a good purpose, but I don't know that we all took it in the spirit for which it was intended. The object, it seemed, was to make for light feet—to be able to surmount obstacles that might arise in combat, and to surmount them without thought. Consequently, every known diabolical road block was put in our path at one time or another. Beyond a doubt, the most difficult section of the course for me was scaling the wall. This was a board wall ten feet high, with convenient places on either side to walk around. I tried walking around the first time, but a curt word from the sergeant changed my mind. He had become accustomed to people trying to walk around. So I went back several steps in order to get a good run at it. The procedure would involve running up, jumping to get a handhold at the top, and then pulling myself up and over. As simple as that.

I cut loose and shot up to the wall, flying spread-eagled through the air. In the first place, I misjudged the distance and jumped about six feet too soon. In the second place, I forgot I was wearing my full field equipment. Consequently, I missed the top and slammed flat against the wall with a resounding smack that could be heard in the next battalion area. Slumping to the ground, my first thought was that I had been castrated.

I picked myself up and quickly stood erect. I was a squad leader, you will remember, and I wanted to look as though I had merely been sizing up the wall for a final attempt. I returned to the take-off point, only not quite so far back this time. I knelt like a panther coiled to spring, leaped up, sprinted forward, soared into the air, reached up and grabbed the top of the wall. A celestial chorus greeted that success with multiple voices. *I made it!*

But no, I hadn't. Not yet. There I hung, clinging with my hands and no promise of getting over. To pull myself directly up was not a feat of which my muscles were capable. Apparently, then, the only way was to swing myself like a pendulum until I had gained enough momentum to hurl one foot over.

If I got one of *my* shoes over, I'd know it was a cinch. So I began to swing, bouncing out and in, banging against the wall, doing the splitz in an effort to get the foot over. I glanced back—a crowd had begun to collect. So with one final supreme effort, I got one leg over and pulled myself to the top. See? Squad leaders can do anything.

Then I looked down. Before this time, I'd boasted that I'd never had any broken bones. Now I realized that period was at an end. I let go and landed in a crumpled heap at the base of the wall. I dragged myself away so that no one would drop on me, and examined my battered limbs and torso. Happily, I was all in one piece. I sprang up, ran cheerfully on to the next part, tripped and fell flat on my face.

One ingenious, diabolical exercise was to walk through a labyrinth of low posts and crossbars without tripping. The next step was to crawl *under* a similar arrangement, presumably without touching posts or crossbars. All designed for soldiers no taller than five feet five inches.

There wasn't any of this hands and knees stuff. It was inch along flat on your stomach like a salamander. Burs, rocks and twigs, scooped up on the way, got channeled down inside the shirt to the stomach. When you reached open air again you raised up in delight and invariably I would raise up too soon. We lost a lot of good men that way. But the sergeant kept us going with words of spirit and encouragement like "Get the hell through there or you'll be gigged the rest of the week!" It spurred us on. Then when we finished, exhausted, and tried to sit down for a well-earned rest, the sergeant called us to attention and double-timed us half a mile. Which helped the burs and twigs to work their way down into our underwear. We were sore in every bone and muscle. From then on I always listened in sympathy to someone complaining about rheumatism.

Extended order drill was another exercise that kept us in scratches. We captured Trench Hill several times by advancing on it tactically, then throwing ourselves to the top to claim the ultimate victory. Extended five yards apart, members of our squad went forward cautiously. When the signal came, we flopped to the ground and crawled along. We hadn't yet learned the proper way to fall, and "any old way" wasn't the thing to do. If we fell flat, we suddenly became breathless. If we angled down, we'd crack a shin or skin an elbow. And we were never in a position to choose a soft clump of grass in which to fall. But we made it, with the lieutenant's compliments as a reward.

One day we went out to observe techniques of concealment and camouflage and it was raining an average Alabama cloudburst that lasted all afternoon. We went from station to station, observed various objects camouflaged and at one point were required to pick out an entire platoon concealed in the meadow before us. When what was over, we had to conceal ourselves in the mud for the next group. We had on raincoats, but mine only reached to the knees, which meant that the leggings formed a trough through which rainwater ran directly into the shoes.

We had a few dry, warm days, when the ground had dried out and become hard as steel, and on these days we practiced digging foxholes. They called it Hasty Field Fortifications, and we began to wonder why so much time was spent on scientifically teaching us how to dig a hole. In any case, my foxhole had to be a foot longer than those of the other members of the company. *Sic transit gloria.*

I had to admit, reluctantly, that beginning basic was good for us. It put us in shape for things to come and taught us a great many techniques upon which some day our lives might depend. By this time we were becoming veteran soldiers. Uncle Sam's best trained, and the sense of accomplishment began to mean something, build up pride, give us a new ego. At least the one compensating feature of the first days was that we could come back to a nice warm hut at day's end and relax on our bunks. In the huts we found life no less exciting and fraught with incidents than training in the field.

Shorty at Hut 9

FIVE
CANDLELIGHT AND KP

Garrison life in the Army wasn't at all half as bad as it would have appeared from the outside. It was regulated, true, and that in itself was a good thing, for the idea of cleanliness and orderliness was paramount. It was a clean place to live, and we kept it that way. The company area was where the developments of personal friendships and loyalties took place. Rough, tough maneuvers in the field threw us together in a compact group. We were the victims who shared the results of the crime now being perpetrated upon us, so to speak. And when we got into the company area, back to our hut, dog tired, and flopped on our beds exhausted, we knew we'd been through something—that we'd been through it together and, having weathered the storm, we came to be a closely knit group.

With barracks discussions, PX sojourns, movies and all the other diversions, it wasn't long before the fact became obvious: we were becoming a coordinated group of combat soldiers whose brainpower had accidentally put them in a radio platoon. Naturally, it wasn't like home (thank heaven), and there were indeed times when we sickened of the atmosphere. On the whole, however, these times were few, for Stonewall had instilled in us from time to time a thought we all harbored only too clearly.

"Brother," he said, "your heart may be at home, but your feet are planted here—so you'd better get used to it."

After a while, we did.

Company life in C-15 wasn't difficult to adapt to. There were three rows of huts pointed north and south. The eastern-most row contained the orderly room and headquarters platoon. The middle huts had message center and the third row contained the wire platoon on the south end and the radio section on the north. There was a fourth row to the west which contained the cadre hut, latrine, and day room.

The mess hall was an east-west building on the south end of the company area. Being up on the north end, as our radio hut was, placed us in a remote position from the mess hall. That was bad. Certain sprinters in our hut, however, became celebrated company chow hounds.

The chief tactical advantage of Hut 9, however, was being right across the street from the latrine. That had crucial benefits. Especially when the food was bad.

The orderly room served as company headquarters. The first sergeant was caged here, as were the captain and some of his staff. This was the mail room, too, and became a popular place at noon and in the evenings. The mail orderly opened the window of his little cell and looked out over the street to call mail deliveries. He delivered them air mail, taking a general aim in the direction of the recipient, and letting fly. During windy days, the letters flew off course and the whole group would fumble around passing letters back, forward and sideways. On rainy days, though, the mail orderly did show some consideration, and would wait for us to walk up and receive mail.

Packages, no matter how well marked FRAGILE, received no special consideration. They, too, were delivered airways. Some weren't too well wrapped and would land in the receiver's arms as a handful of cookie crumbs. As the mail man put it when he carefully passed out a dilapidated package: "Another five feet and that one wouldn't have made it!"

A bulletin board at the south end of the Orderly Room carried notices, training schedules, and the weekly guard and KP roster. We often found it difficult to look at this board. The day room was a popular place from time to time. Here were writing desks, occupied most of the time, even though 90% of company letter writing was done in the huts. The day room had a ping pong table and a larger table for card games. One corner was kept clear so that the floor and walls could be utilized in the African dominoes pastime (throwing the dice). A soft-drink machine added revenue to the company treasury. We signed the payroll and were paid here. It was a room where a lot of money was lost, too. Besides dice, poker games were frequent during the week following pay day (after which other less expensive pursuits took place). By and large, however, the troops played poker mostly in the huts.

As many as eight players huddled around one bunk on which the cover had been pulled tight and served quite adequately as a playing board. On days off, these games usually persisted throughout the day, with the participants getting up once or twice to stretch, or make a last-minute dash to the latrine.

The sounds of such games were characterized by brief, periods of heavy breathing and meditation, broken by bursts of irate voices.

"Son-of-a-BITCH!"

"You BASTARD! I thought you were *bluffing!*"

"Jesus CHRIST! I never saw such a lucky streak in my life."

"Jealous! You guys are just jealous."

"I'm also broke. Count me out."

"Okay. The rest of you ready?"

"Deal, goddammit! You ain't gonna last forever. I'll skin your *ass!*"

"Haw. I'll clean you out."

"Fuck you! DEAL!"

A common sight was two poker games in the same hut, but they usually differed in the wagering media. On one bed lay one and five dollar bills, on the other nickels and dimes. There was another difference. Those whose eyes peered furtively over the folding money had worried looks. Those who gathered about the clinking coins enjoyed the game. Whenever I played poker, I enjoyed the game.

Quite often, on a Sunday afternoon, with a cold rain pouring outside, the fire was stoked up, the doors closed, and the windows shut tightly. The result was a muggy, stuffy, stifling atmosphere unnoticed by the players intent on their cards. It was just the kind of pregnant atmosphere Winnie Morgan waited for. Waited until a particularly tense moment of a game, waited until men were clustered around, bent forward attentively, watching or playing, and then cut loose with a long extended, hollow-voiced, boomlike fart.

The crowd reeled back. Players sprang up and opened the windows, uttering epithets. Others hurled themselves out the doors.

"Winnie! You BASTARD!"

"You son-of-a-bitch! Somebody bring another pair of pants for this asshole!"

Winnie reveled in his role as champion farter of Hut 9. The soldier's obsession with excessive flatulence in those days was such that anyone, any time, could cut loose unashamedly from anywhere in the hut, and send his bunk mates scrambling for the exits. But somehow, biology had endowed Winnie with the rare capacity to store up ammunition, as it were, and wait until the best of all moments for release. He even jumped up on a foot locker, once, and produced a soprano aria that made him famous throughout the company. A legend in his own time.

Above all other facilities, the latrine had to be kept spotlessly clean. Ordinarily, it was large enough to accommodate the morning and evening rush hours. Here, too, were held bull sessions—but then, the atmosphere suited that type of activity.

Still, it was the huts themselves that meant the most to us, apart from Winnie's transgressions. They were about 25 feet long and 12 wide. A main row of beds were placed side by side down one side and another row end to end in the remaining space on the other side. With foot lockers at one end and barracks bags at the other, there didn't happen to be a great deal of room. As a matter of fact, scarcely six inches separated bunks. We had to laugh that SOP (Standard Operating Procedure) called for 18 inches between bunks, for health and happiness. And yet we got along all right, though at morning we had to make our beds in shifts or take the consequences.

We each had a section of a shelf that ran the length of the building, except for the windows. Our clothes hung beneath each shelf. The shelf of each individual contained standard equipment such as picture, canteen, and shaving materials. During the early part of the cycle, a pot-bellied stove in the south end of the hut kept most of us warm. When it burned at a heat which those nearest to it could stand, those of us unfortunate enough to be in the middle of the hut sometimes stayed cool. But for the most part, we kept dry. Later, the stove was taken out and we reveled in the acres more room it made.

Sunday mornings were usually the only times that absolute quiet reigned in the hut. The rest of the time was given to varying activity. If two men at each end of the hut wanted to carry on a conversation, they did so without moving, at least as best they could above the noises in between. The general sounds were those such as clink of rifle parts, scratch of pens, whisk of shoe brushes, and scrape of hand brooms on clothes. When lights went out each evening at nine, 12 or 15 candles were lit and activities continued as though nothing had happened. I know that the greater majority of my letters were written astraddle my foot locker with my hands and arms wrapped around a candle. It was almost always to bed by ten.

The people randomly gathered together in circumstances like these offered an interesting cross-section of the country that sent them to war. The greater majority of our company was made up of recent high school graduates, college and business men, and those who had studied music, divinity, the fine arts, and sciences. Hut 9 had its share of this microcosm.

Gil Bower, 28, short, round faced, and good natured, had been in the CCC (Civilian Conservation Corps) some ten years before. He would have made an excellent efficiency expert for he always got things done early and was invariably the first one in the hut to hit the sack each evening, reading his Bible before going to sleep. He sang in the chapel choir and once got me interested, though I protested that I didn't have a voice, and sang a refrain to prove it. Nevertheless, he dragged me up to one or two rehearsals and I thought I might like the experience until things got too busy at the company. It was difficult to maintain a choir on this basis, since there were so many absences from influenza and pneumonia. Yet each Sunday at church, they did a commendable job. Bower's home was in California, and he'd come out from MacArthur on the same train I did. We often wondered why he never got much mail, but that was none of our business. We never, never pried into the private lives of our colleagues. We usually didn't have to. Most of them were so voluble that we soon knew all about them anyway.

Bob Hill had come out from MacArthur, too. In fact, I first met him in one of the numerous lines back there and we palled around some. We struck up a good acquaintance and became fast friends at McClellan. He was small, light and white, but stood up under the rigors of basic as well as the best. Bob was from Hollywood and promised me a personal tour of the film capital should I ever come out for a screen test. I think some day I'll take him up on it. He knew where Betty Grable's house was.

Jimmy Ruff was from South Dakota and he and Bob Hill and I came within a hair's breadth of being triplets. At the time, we were all eighteen, and our birthdays ran consecutively the 9th, 10th, and 11th of October. Being the oldest, I had to give them fatherly advice now and then.

We had perhaps more than our rightful share of Iowans. Winston Morgan, of whom we have already heard, was 23, a handsome kid, glib and sly at poker and always springing some cutting remark fresh from the blade. He'd been a night-club singer and welder back in California before the war.

Other Iowans were Andy Hochstetter, whom we called Hockey, and Hency Gattis, both about 20 and both apparently from high born families. Hockey especially had well-trained home connections that insured his receiving a box about every other day, and of course when he, or any of us, got a box from home, it immediately became Hut 9 community property. That made it a little difficult in writing home to the folks and asking for, say, a cake. You had to suggest that it be big enough to be cut twenty ways or you wouldn't get a piece of your own. We were always free with our heaven-sent gifts and most of the time you could walk down the aisle of the hut and find a box of candy or cookies lying open on a bunk. And you just reached in and helped yourself. Asking for a piece was not considered good taste though you must express thanks some time during the day. Thus, we seldom went hungry.

"Said something about Iowa."

Scott Foss, 19, ruggedly-built and likeable, seemed well versed in many subjects and could take jokes very well. He was from Minnesota.

Kid Harrick was another young Iowan who seemed at first to be one of our incurable club-footers but soon caught on and when he did he was a first rate GI. We were always saying he was dumb and blissful and we always got a laugh out of him, which is one big reason why we all liked him. In fact, so friendly were we all, that insults came to be badges of honor.

One night, the Kid had just been across the street in the latrine washing his mess kit and that of a friend, and we could hear him stomping back to the hut, mess kits clanking, and he was humming some corn country air. It was shortly after lights out and Gil Bower, as usual, was asleep. When Harrick came in the door, he tripped on the steps and lost his mess kits, which came entirely apart and went clattering and slamming from one end of the hut to the other. Bower flew straight up in bed and barked, "What was that?!"

Harrick split the air with a blue streak of corn-country cussing, picked up the parts of his mess gear, and went out the door to rewash them, blissfully humming a tune. We heard him stomping across the street and finally he, his bliss, and his music disappeared in the latrine. I believe being like that has its advantages. He could be proud—he was our hut comic.

Evan Marsh was the hut character and chief poker hand. He was 24 and from California. His dislike of Army life kept always fresh, but the way he cussed it somehow never reflected on the Army, and you could tell he was having a good time.

There were two other Arizonans in the crowd: Perry Perrine and Al Coggins, both about 28. We called Coggins "Gramps" for he was the chief gripe artist and made most of his attacks on the rough training, which he said made him grow old before his time. He got his copy of *The Arizona Republic* each day, and if I ever had a free split-second, I'd grab a copy and read some of it.

Gramps, by the way, was also a motion picture projectionist and had worked the same chain I had, though he had lived, I think, in Kingman. So I had help when the battle of the States took place. Iowa and Arizona particularly got hell from both sides of the hut, but I think we three Arizonans won many of the battles by fishing out a copy of *Arizona Highways* and letting the Iowans have a peep. They cooled down when they began to realize ours wasn't all a desert state. But sometimes when the battle waxed hot, Foss would break in to say that when any of us wanted to see a real God's country, why just come on up Minnesota way. So we all broke up to run him out of the hut.

The oldest member of the hut squad was 37-year-old Burl Matthews. He came to known as "Lieutenant" Matthews after a boner he pulled one day in a classroom, something about standing up when the instructor called to an officer in the back of the room.

Aubie Darr was a typical high school lover and playboy, and he and Matthews were both Kansas sons. Chuck Griggs, a hardy lumberjack from Montana, had one of the best personalities in the hut. Sam McCurdy, from Colorado, kept getting sick, and it amazed us how he could miss so many formations and still be alive. He loved the wine, women and song in Anniston, and so had the challenge of straightening out his priorities.

Walt Garrison was a good-natured kid from California. He'd once studied in a seminary, though that experience couldn't have been guessed. Bower was the only one who didn't give vent verbally to his feelings, and we sometimes wondered what he thought about the rest of us bums. We didn't worry, though. When a situation called for some proper remarks, the remarks got made. All in all, Hut 9 had a good bunch, and we all got along perfectly throughout training. One big happy family...

Another building in the company area which received considerable attention each day was the mess hall. When the chow whistle blew, all thought of aching bones and sore feet was forgotten in the mad dash to get in line. Bower, however, was the fastest and somehow I managed to be next. If that didn't make us chow hounds, the daily trips for seconds did. We at least always managed to leave well filled.

Eating was family-cafeteria style, in which we'd pick up the main courses just inside the door, and garnish the plates from the tables. The food was quite all right, well balanced and well prepared. Most of the time there was enough, but later the CO (Commanding Officer) managed to get us more rations, and we thrived on the great abundance then. We had the best baker in McClellan, and some of the pastries he put on the table in mass production rivaled those prepared with care and delicacy back home. The only trouble was that we were seldom permitted to go back for fourths and fifths. Bower and I had to be frequently turned down. Standing in line for chow was disagreeable, especially when it rained every day. Finally someone wised up, and we ate by platoons, which reduced the competition and saved considerable effort and racing.

The kitchen, however, was not the most pleasant place in the company on KP days, and not even squad leaders were exempt from this phase of training—much to my chagrin. The disagreeable aspect of the day began when we had to drag ourselves out of bed at 4:30 a.m.—sometimes made worse when the night before had been spent on a tactical exercise that lasted until midnight. That was no excuse, though, because if we weren't out in the field, we'd be at the movies or writing letters or some other place we had no sense to be if we were going to get up so early.

Duties involved everything from A to Z—serving, waiting tables, washing and scrubbing floors, cutting bread, slicing butter, and that old standby, potato peeling. We normally got off a couple hours in the morning and afternoon, and were through by eight o'clock in the evening. If we'd done a good job, the baker would reward us by dipping some cake out of his secret cupboard and treating us.

The most unforgivable sin that could be committed against us was to have KP on Sunday. That was the one day in the week which we all relegated to catching up on such things as loafing and letter writing. KP on this day meant additionally that we weren't escaping drill, calisthenics and other unfavorable rigors of basic training. I had it one Sunday and had to work three hours overtime in the morning to boot, got half an hour off, then went back for dinner service (dinner being the noon meal in the military lexicon and supper being the evening meal).

Taking off prematurely for the afternoon, I went to the show with some of the boys, then went over to visit some fellows in another company. As it turned out, supper was two hours early that evening, so I arrived late. Consequently, I was put to work cleaning out the three massive kitchen stoves and carrying in coal. It wasn't the cleanest job in the Army, and I wound up sooted from head to foot. But then, punishment be to the wicked.

Guard duty wasn't popular, either, but at least being a squad leader placed me in the position as Corporal of the Guard, and I didn't have to go out and stand duty. This I particularly enjoyed during the rainy season. Tromping out to relieve the guard wasn't half as bad as standing in the stuff for hours.

And perhaps it was just as well, for my raincoat, besides not being long enough, began to come apart at the seams. It apparently had been glued together, with a glue that wasn't strictly waterproof. So the Alabama dew sometimes seeped in. But that still didn't mean anything, for the others came back in from guard duty soaked to the skin, shivering and cussing. And I found, after many hours of rain in the field and on the firing range, that no matter how good the raincoat, the helmet makes a poor umbrella and no one can satisfactorily keep water from seeping in—first along the neck, then down the back and front, until at last you feel as if you might be drowning. Sloshing back to the huts in formation sounded like a lot of cows stuck in the mud.

Latrine duty was another inescapable detail that crept into our private lives, though not nearly as often as guard and KP, for the latrine was not in use when the company was gone and one man could usually have sufficient time during the morning to take care of it. When inspections took place, various squads in the company would take turns spending a half hour there getting the worst of it cleaned up before they had to fall out, then the latrine orderly finished the job and hid somewhere until after inspection.

One of the greatest hazards to the serviceman's mental well-being is the problem of personal inspections. Sure, everything had to be clean and shining and there were a number of pleasant ways to force us to do it. But the Army did not select any of these methods. There was a continual cracking of the whip and at each inspection the axe would fall here and there like lightning striking in a heavy thunderstorm. And the storm fell on us a number of times.

Most of the time we were expected to perform miracles of cleaning and preparation. And in the shortest possible time. We didn't complain, though, not after the first time. We took our sob story to the sergeant and he said: "Well, you know the old Army saying—the difficult we do immediately, the impossible takes a little longer." So we knew that further crying on the shoulder would bring no relief. Maybe the principle was good...ask for the good, get a fair job; ask for the impossible, get a good job. At any rate, we tried to keep busy and out of mischief.

One night we came in after a rigorous day in the field and had about two hours before lights out to get everything in order for a big inspection on the morrow. Two hours in which to clean rifles and bayonets, arrange lockers, shower, shave, and do laundry. Then one of those precious hours was utilized in drawing some equipment. We at the end of the alphabet stood in line for an hour. That meant that a great many things had to be accomplished after lights out. Flickering candles moved all over the hut.

The following morning I took a brief moment to shine shoes and put the finishing touches on arrangement of objects on the shelf. It would be an indoor inspection—of us and all our equipment.

Perhaps the worst part of an inspection of this type is the standing around waiting for the inspecting officer to come in. We would pace around white-faced, brush a speck of dust off the bed, turn a picture a little more to the front, rearrange the hair oil and shave lotion to different places in the foot locker—anything to keep busy in those last moments. By peering furtively out of the windows, we could see which hut was being inspected. When the inspecting party reached the next hut, we all got in position and readied everything in a last flurry of nervousness—pressed a fold out of trousers, combed the hair, lined up a shoe.

Finally, a breathless hush fell over us as the door of the adjacent hut opened. Footfalls down the steps, across the walkway, up our steps. We all looked straight ahead. Hearts pounded. The door opened and there was a loud call to attention. Then more footsteps. In the upper end of the hut, which I couldn't quite see out of the corner of my eye, I heard low, ominous mutterings—and the deadly scratchings of a pencil. As the officer came closer, gigging right and left, I wanted to crawl into the nearest small hole and hide. Then, at last, he stepped in front of me.

I could feel my skin crawling. *Who do you think you are? I wanted to ask. This is a godammed free country. If I want my shaving lotion out of place, I can damn well* have *it out of place, and it is none of your...*

I could feel my skin crawling, my throat suddenly dry. I was being looked over, but good. And what made it so bad was that I knew I wouldn't pass, and I could think of a thousand reasons why.

"Dirty screw heads," came the low monotone, referring to my rifle, then the scratching of the pencil, then on to the next bunk.

Dirty screw you! I whispered to myself. *Up your...*

"Dirty bayonet." And so it was all along the line. Virtually everyone was marked down for some infraction, all because of the universal excuse: NO TIME. But then, excuses didn't mean anything.

However, not much came of this inspection. Whenever *all* the men have something wrong, the officers grin wryly to themselves and mutter, "Well, maybe the men *should* have had more time." So we got passes and all normal privileges anyway.

I take it back, I muttered to myself.

These inspections were held regularly every Saturday morning. If we passed, okay. If not, some ingenious method of penal servitude would be drawn from the files. Occasionally a rifle inspection would be held at retreat (the evening assembly). When this happened once, someone left the rifle rack unlocked and the lights on in our hut. This was discovered during our absence at the inspection, and the tongue lashing we got gave us to know that this negligence of duty just about equalled a horrible old court martial offense. Therefore, over the weekend the members of the hut were lined up and marched over to report to the orderly room every two hours. After that twenty men individually checked the rifle rack and lights every time we fell out for the rest of the cycle.

In going from our huts to the parade ground south of the company area—at which place retreat was normally held—there was a ditch that was crossed by a small footbridge. Now when 200 men rush out to the parade ground at once, only about ten who happen to be in the right place receive the privilege of crossing the bridge. On one sunny evening, we were rushing out for a retreat inspection. I had meticulously cleaned my rifle in the hour or so beforehand, and cleaned it to exacting SOP (Standard Operating Procedure) standards. Take apart, wash with soap, dry with rifle patches, oil, put together—and for once, I had it clean. And then, of course…

I plunged down into the ditch all right, but up the other side I got off balance, and three steps beyond I tripped and fell, rifle, pressed uniform and all, through six feet of sand before coming to a stop. This wouldn't have been so bad except that I had forgotten to close the chamber of the rifle and when I got up I poured a cupful of sand out of it.

I stood inspection, there being no other choice. When the captain saw my weapon, I thought his jaw would drop out of its socket. Fortunately, he took Stonewall's word that I'd just fallen in the sand. After all, *nobody's* rifle could be that dirty. Not even mine.

The procedure at inspection arms was to bring the rifle up and snap open the chamber in one sharp movement. Then, as we were so carefully instructed, when the officer comes to take it, almost as soon as his hand leaves his side you must drop the rifle. From then on, it's his, to have and to hold. I remember once doing this, just as they said. And the officer *missed* it. It dropped barrel first into the ground and he caught it just in time to avert a complete catastrophe. For to think what happens when we the rookies drop a gun is foreboding enough. What a horrible embarrassment it might be to *him* to drop a gun. And what a worse embarrassment to me to be the instrument of his misfortune. His only comment was: "You really let go of that, don't you, bud?" We all held back our riotous laughter until we got back in the hut.

There was another time, though, when things weren't quite so rosy. It happened during a lull moment at retreat, just before the regimental ready signal was blown on some distant bugle. We stood there spic and span, each the epitome of the perfect soldier. Everything was quiet. Nobody said a word. We were at parade rest position.

"Pretty sharp-looking company, don't you think?"

The captain said, in a low voice, "Well, men, how do you feel after the session today?" There were a few nods, meant to convey an "okay."

Then he said: "Where's your tie, Sutton?"

I don't believe I'll ever experience a predicament as embarrassing as that again. All the other nattily dressed members of the company turned and stared at me. Me, the squad leader. Me, the paragon of virtue and efficiency.

My hand flew up to my throat as if a noose had just been pulled tightly around it. Sure as blazes, I had forgotten to put on my tie, and in the rush to formation, nobody else had noticed it and warned me. I was stunned. I forget what I muttered in answer. The worst part was standing in the front row, directly in front of the captain. That's what I got for being so tall... shoved right up under the commanding officer's eyes.

Just then, the bugle blew and we were called to attention for retreat. Afterward, Stonewall gave me one of the coldest stares I've ever received.

But the captain was a good guy and forgot it. Unfortunately, he was later replaced as our CO and we were all extremely sorry to see him go. The captain who took his place was merely a playboy in uniform who utilized his powers of tyranny to the fullest allowable extent. He drew no respect from the men at any time. He completely washed himself up with us when he once led us in a four mile double time and had to fall out to catch his breath while we went on. That wasn't good.

Reveille and retreat were as regular as clockwork. Once in a while, as we stood there with freshly shined rifles and newly pressed uniforms, and it began to rain, we were extended the pleasant courtesy of being dismissed to run pell mell back to safety. We could never fully appreciate reveille, since it came fifteen minutes after we were awakened. But retreat was always impressive, at least from the standpoint of hearing the distant regiment bugle and imagining the ceremony and the lowering of the flag. It wasn't flag-waving and gave some of us a little renewed national spirit and patriotism after a hard day's labor. I was never ashamed of being patriotic. I could cuss out my country sometimes, but that's why I was patriotic.

Parades were always impressive. When the entire regiment marched, row on row, bayonets bristling, it was an eyeful from all angles. Of course, I'm sure I'd have enjoyed the view from the viewing stand just as much. But then, the war required that expert soldiers be in the ranks, too, so I stayed where I was.

We got our laundry done each week at a price of $1.50 per month. It would seem that during all the muddy weather, the launderers would have lost money, but they did a good job. Half the time, though, they got things mixed up and few of us could retain our original socks and underwear throughout the cycle. I was lucky they never accidentally exchanged my trousers with anyone else. I'd have been sunk. Knee britches are frowned upon as GI apparel.

We aired beds every Thursday. They were placed in the company street and the blankets draped over them so as to receive full sunlight. This practice, however, had to be limited to dry, warm days. This gave us a chance to clean the barracks out stem to stern, though the most thorough cleaning was frequently done Friday night, prior to Saturday's big inspection. In the latter case, the beds were hauled outdoors along with footlockers, barracks bags, and clothes. Then the hose was brought in, and the hut was sprayed mercilessly. A brigade of brushes and mops followed, and by the time the smell of fresh lumber came through, we quit and hauled all our belongings back inside. The windows were thoroughly scrubbed, too, often with a vehemence. One of the boys once broke a pane by rubbing it so hard. But that was the way we passed inspections.

They told us early in the cycle that if we were good, that is to say, if there was no AWOL (Absence Without Leave) or VD (Venereal Disease) in the battalion during training, we would receive eight days furlough instead of the customary seven. Somehow, I knew we'd get eight, but then I should have known better. Anniston was just too close to camp. Just the same, it seemed cruel to punish that many hundred men for the misdeeds of one or two, yet that type of treatment persisted for a long time in the Army.

When we had free evenings, time was spent largely in cleaning rifles, rolling and rerolling packs, shining shoes, and letter writing—which always lasted late into candlelight. On such pleasant evenings, bull sessions went on vigorously. Then a couple hours after chow, hunger would set in and someone would ask who was going to the PX. No one would answer. Finally, after several coin flips, the unfortunate would collect orders and money and take off, returning under a staggering load. One night when I did it, this was the list:

PERRY .... 1 Chesterfields or Camels
HENCY .... 2 candy, 1 pint chocolate ice cream
BURL .... 1 pint vanilla
JIMMY .... 1 pint chocolate
WINNIE .... 2 Camels
CHUCK.......... 2 Chesterfields, 1 notebook
WALT .... 1 Camels
SUT .... 1 Brillo, 5 Camels, 2 candy, 1 notebook, 2 pints vanilla

If we were this hungry at night, you'd think the Army never fed us. On the contrary, the Army must have lost money on me when it came to chow. A typical breakfast consisted of prunes, 3 bowls of cereal, 2 hotcakes, bacon, butter, and 2 cups of milk. Dinner and supper were equally devoured. Naturally, with all the exercise we had, the mess hall had to keep up a good stock.

One night, Bob Hill and I went over to Co.A, 9th Bn., to visit a friend we'd known back in MacArthur days. Strictly speaking, he wasn't as happy as we were, and he fully realized how lucky we were to be in a specialists' regiment. His battalion was infantry, all the way through, and while we were only getting seven weeks basic, his entire 17 weeks was a continuous matter of firing, obstacle course, tactical maneuvers and all the other aspects of combat training. His was a rifle company, and he said they'd lost a number of men because the training was so rough. However, he admitted, it was better on the whole to separate the men from the boys there rather than have the enemy do it overseas. At any rate, we gave him our sympathy and best wishes, and didn't talk very much about the radio aspects of our life to come.

Some girl in Iowa wrote to one of the guys in the hut saying she'd seen his name in the paper and wanted to write to some soldiers and would we please oblige. We did, about ten of us, listing our dimensions, home towns, likes and dislikes. She wrote back to each in turn, but the only really important thing she said was: "Myron, you said that you were 6' 5" tall, with red hair and brown eyes. You sound just like my dream man!"

This little episode netted me a new nickname, as if the others weren't already enough. I was known as "Dream Man" for weeks to come.

One of the best Sundays I had was enjoyed in the pure delight of loafing. Since I'd cleaned everything for inspection the previous day and had little or nothing to do at camp, I thought about getting out for the day. Bob Hill and I left the hut at ten and went down to the Main PX, dropping off on the way at A-9 and chatting a bit with our friend. At the PX Bob ate breakfast and I had my second one for the morning. We did some shopping, stopped at the fountain for ice cream, then caught the bus out to the golf driving range. Here we each got a bucket of 40 golf balls and the use of any club we desired—price 15¢—and whiled away an hour with practice shots. We

came back to the Service Club then and lounged in the music room writing letters. Sitting there listening to the classics was spoiling us, we knew, but it was nice anyway. After that, we went over to the bowling alley, bowled a line, then had supper in the PX cafeteria.

I drew out my most beautiful smile and dropped in at the photographer's. We got back home at seven and I decided I'd do something a little constructive for the day, so I rearranged my foot-locker, rerolled my pack, cleaned my mess kit, and did some laundry—after which came the reading of a copy of *Arizona Highways* until lights out, then a letter home by candlelight. The only trouble with a day like that was that there was only one in seven like it. And even that wasn't guaranteed.

Frequently, though, a soft day like that could leave us tired at night. On the other hand a rugged day in the field could bring us in rarin' to go out for the night. Let it not here be construed that all such nights were that way, for other times brought us in ready to drop dead on the bunks.

One such day was when calisthenics began to get really rough. We'd double time around and around the field, then engage in all sorts of new and unimaginable exercises, including pushups. Then another hour of the same thing in the afternoon. Sprinkled liberally in between was bayonet practice. In its advanced stages, it included a position known as long thrust parry right, at which the rifle was held out in a forward position solely by the last glimmering strength left in the arms. Coming in after a day like that, we flopped on our cots and thought we'd never move again. This, however, was only superficial. As soon as the mail whistle blew, the hut exploded and we all rushed out at breakneck speed.

Such was barracks life at McClellan and in spite of its candlelight and KP, its guard and latrine duty, its inspections and cleaning, we did have some good times and got along right well. Our temporary tap roots were there in Hut 9, and as the cycle progressed, that hut began to be looked upon each evening as a pretty wonderful place to get back to. There were other places we'd have liked to get back to, but since the war was still going on, we felt we ought to complete our training so as to get in and finish the job overseas. Naturally, once the Germans and Japanese knew we were coming, they would give up the ghost and quit. So with that aim in mind, basic training roared on and we headed for the firing lines.

SIX
FIRING

On Leap Year Day in 1944 it rained as usual and that night turned cold, so that next day, March came in like an icicle. Here we began the hand grenade phase of training, and it started off with dummy grenades. We were placed behind an embankment, learned our positions and instructions, and rehearsed. *Kneel! Pull pin! Prepare to throw! THROW!* Then we would go out, chase the dummy, bring it back and do the same thing over. By the time this became monotonous—practice virtually every day—we went out on the range for the real McCoy.

After chow on the morning of March 11 we donned our full field pack, cartridge belt, rifle and bayonet, canteen and steel helmet and hiked four miles to a section of Defendam Range for the last hand grenade practice. Just to get back into shape, we spent some time throwing dummies, then had an hour class on the makings of a grenade. After this, half the platoon went up on the firing line while the rest of us withdrew up a hill to watch proceedings through Army 7X50 binoculars.

Each man on the line threw two grenades from a trench. The grenade was thrown, the line ducked out of sight, the explosions came and the report followed a second later.

When our turn came, I was one of the last to throw. Being on the tail end, several of us got to hurl some leftover grenades. It was our first contact with live ammunition, which, for us, turned out to be an interesting day. Throwing grenades in combat, however, possessed an entirely different atmosphere. One thought that came to us during the day was that this would be a most efficient means by which the enemy could knock out our radio dugouts. Just sneak up and lob a grenade in. Radio was a *safe* occupation? In warfare, nothing is "safe." There were more pleasant thoughts during the day, and we hastily turned to them.

On the 6th of March came our first rifle practice, getting the sight picture or marking the target properly. Prior to this, we had received much classroom instruction on the workings of the rifle. The M-1, we learned, was operated on the principle of gas pressure resulting from the firing of the shell. Its muzzle velocity measured approximately 2,700 feet per second. Stoppage was a condition whereby the piece failed to fire and could come about by any number of causes, all of which we were supposed to know and prevent.

"How was grenade practice today?"

The principal means of failure was dirt, and many hours were spent in telling us how to clean the rifle, how to take it apart and put it back together, and how to reclean it when it failed to pass inspection. The whole business was so detailed that frequent inspections were held whereby each individual piece was laid out on the bed in the hut—open and ready to be gigged.

But to this there was a reason. They said the rifle would become our best friend in combat and must therefore be treated as such. There were times, however, when I felt like wrapping my best friend around the nearest tree.

Rifle marksmanship Cal. .30 was interrupted by training in cover and concealment techniques—practice in creeping, crawling, going under barbed wire, running, rising, falling, climbing over walls, up hills, into ravines, and other restful devices for adding variety. Then came training in the different positions from which the rifle might be fired: prone, sitting, kneeling, standing, squatting and anti-aircraft. This was known as dry firing and dummy cartridges were used, eight shells to a clip.

One of the worst times I had was trying to get my six feet five into all those positions. Only a Yogi contortionist could have fit them all and still held a gun. I could get into the positions well enough, but my arms were so askew that I couldn't fire the rifle. I would squat, stoop and stretch, finally squeezing one muscle over another and bending another bone double to find the correct stance. When I looked down the sights, the target lay several hundred yards to one side or the other and I had to get up and try it all over. If I had to depend on the enemy standing still while I got into the right position over in combat, then they could start marking off my insurance then and there.

Though the firing positions never did become natural, I got used to some of them and was at last ready to fire. One of the rules that governed our personal appearance at all times was that of keeping pockets buttoned. Once on the practice range, during squat and bend practice, one of my pockets came unbuttoned mysteriously and presented itself to a passing lieutenant.

"Sutton, report to me at the next break."

"Yes, sir."

At the break, I found Jimmy Ruff in the same predicament, so we set out together, half-heartedly, to receive our 20 lashes. Except that the punishment here was worse—five to fifteen pushups, which, for me, were never a cinch. We found the lieutenant all right, but he was talking to some other officers and we figured it would be discourteous to go up and bother him with this trivial matter. Perhaps he'd forgotten about it. Convinced that this might be the case, we straggled back into the shadows and sneaked away. This, it turned out, was a mistake.

Back on the firing line, the lieutenant saw me and said: "Sutton, weren't you supposed to report to me during that last break?"

Which was a ridiculous way to put it. *Hell yes, I was supposed to report to you during the last break—I ain't deaf.*

I said: "Yes, sir, but I couldn't find you. When I did, you were talking over there."

"No," he replied. "I was on the line here. Boy, you'd better learn to obey orders. Who's your platoon sergeant?"

"Sergeant Stone, sir," came my weak reply.

When the rest of the platoon fell out for chow at noon, Stonewall called Ruff and me aside.

"You were ordered to report to the Lieutenant, weren't you?"

We nodded.

"And you didn't do it, did you?"

We nodded again, heads hanging lower.

"You know what that means, don't you?"

Weak nods. I asked plaintively, naively. if there was any way to get out of the presumed punishment to come. Stone replied: "You have about as much chance of that as a fart in a hurricane."

Our heads drooped further.

"Well, I was going to do something about it," he went on, "but you're still a little new, so I'll leave it up to him. Report to him at the first break this afternoon."

"If you weren't a private I'd break you!"

We fully expected this to continue all day, and it very well could have. In the afternoon we reported to the Lieutenant and stood there listening to a talk on discipline. It was learning the hard way, but it was learning. Punishment: five humiliating pushups.

"Couldn't you see you were diggin' the damn foxhole in an anthill?"

The last of our training in hasty field fortifications served to show us the protection offered by the foxhole. Our foxhole, while we crouched at the bottom of it, was run over by a 14-ton light tank, an experience not dissimilar to that of an earthquake. Afterwards, we were shown the makings of the tank and crawled all over it in close scrutiny. Here, too, came an introduction to high explosives when we learned about anti-tank mines, bazookas, and the Molotov cocktail fire grenades. March 13 was the first live-firing day. I was on ammo detail that morning and expected there would be a lot of weightlifting (boxes of shells are *heavy*). As it turned out, we loaded only a few small items and received the extreme advantage of riding out to the range. The rest of the morning was spent in passing out ammo when it was needed on the firing line.

This took place at McClellan's Washington Rifle Range, limited to the small arms field. After dinner, our platoon fired.

The first stage was a ten-inch target from 200 yards in order to adjust proper elevation and windage on the rifle's sights. Then we moved back to 300 yards and made more adjustments.

This wasn't so bad. At least, I didn't draw any Maggie's drawers (red flags raised when the target is missed completely). From there, we moved back to the 500 yard line and I asked where they'd taken the targets. Then someone pointed them out to me, and by squinting, I was able to make them out. I thought surely they must be clear over in the next county. Here again, I struck the target every time, even made a bullseye—but then, accidents will happen. In this zeroing-in process I managed to break 50 points out of 60. At that rate I knew the expert medal would be forthcoming in a short time. They were offering a $25 bond as first prize for the highest score and I was already figuring on mailing the bond home.

After every round fired, we opened the bolt, locked it and held the rifle on our shoulder to be cleared by an officer. Each rifle was inspected for emptiness again at the end of each day. It was a court martial offense to carry live ammunition or fired empty shells off the range.

By now, an Alabama spring began to seep through the winter cold. The rain lessened and the grass started coming out in full scale green. Flowers appeared in the fields and on the trees and the sun at last grew warm. Wool undershirts worn in the mornings came off at noon. As a matter of fact, it got downright hot one day, so that a cold shower that evening actually felt good. This was a preview of the sweltering days to come.

And yet winter hadn't had its last swipe at us and the warm days that came were interspersed with torrential downpours and shivering cold. At any rate, it felt good to work up a first-rate honest sweat again.

We finished bayonet training by charging dummies and using different parries and thrusts to ram them. Some of the boys were in good spirit and a number of bayonets were broken in this way.

Another night we had to have our rifles cleaned and laid apart on the bed just because two or three men couldn't keep in step coming home from the range. The powers that be grasped at any excuse to get us to clean those rifles.

As rifle training continued and the cycle advanced, there seemed to be more and more work poured on us. Noon mail call was discontinued, and that caused no end of griping. It meant that there was such a load of mail in the evening that an hour was required to pass it all out. A precious hour of time. On several occasions our lunch hour was cut short—down to 20 minutes for the consumption of a big meal. During this period, we made heavy runs on the PX supplies of Tums and Alka Seltzer.

Noon mail call, Fort McClellan, Albama, 1944

There seemed to be a never-ending parade of shots to be taken, and those at the end of the alphabet always came straggling in last. Then, just before retreat, we'd be told—again—to clean our rifles. Since they'd been sitting in a drizzle all morning, they were as rusty as a rainbarrel. As a rule, I didn't even get started cleaning mine by the time we were called out for some physical training. After that, we took off everything (in the hut, of course, where do you think?) and put on helmet liners and raincoats. Following this inspection (tongue out, arms up, breathe deeply, cough) we would return to the company area to get into retreat uniforms. And there was the rifle, still rusty. With a total of five minutes left, I dared not take the weapon apart to clean it—so I sort of brushed here and there in the most obvious places so that it wouldn't smart the captain's eyes when he saw it.

After inspection, which is to say, after gigging everyone, they told us severally that as a whole, the rifles were a mess, a fact of which we were only too well aware. And I have a sneaking suspicion they expected it. This didn't happen too many times before we began to get more cleaning time before inspections.

Back on the firing line, a little more than half of us qualified (made a specified score) the first time around in firing the M-1. While the others tried again, we who had passed were assigned to helping them along, or, as I was one day lucky enough to achieve, assigned to guarding the platoon packs. This meant hours of sitting, relaxing or catching up on writing a letter, looking over on the firing line and watching the rest of them slave away. Some passing troops muttered about my being a lucky dog or some such phrase. I thumbed my nose at them. Really, if you weren't an expert rifleman—if you weren't going to be a good soldier—then you actually couldn't expect rewards like this. Naturally, such a haven of rest, insignificant as it was amid the hurry and scurry, brought on accusations of another process known variously but seldom by its polite form—polishing the apple. But then everyone got called Brownie when shown the least courtesy from the sergeant, or when volunteering for some detail.

Next came firing the rifle grenade. And they never told us what to expect. The best position was to place the butt of the rifle on the ground and fire.

"Was there much kick?"

But we were theoretically tough enough, we thought, to take it from the shoulder. So that was where we put the butt of the rifle. One grenade was all we were allowed to fire, and one grenade was all we cared to fire. *POW!* I was positive I'd dislocated my shoulder, and examined the bones carefully after picking myself up off the ground. Firing that grenade, I felt, would do more damage to the firer than to the enemy, but then, never having been hit by a rifle grenade, I wasn't in an authentic position to know. I would fire it from the ground from now on.

The day we started firing the M-1 rifle for record was a lovely day: dark clouds all over the sky, an ice cold wind, drizzling rain. This was the atmosphere in which we must fire our best. I did okay from 200 and 300 yards but from the 500 yard line the weather really played havoc. At least, that's the handiest excuse I had for missing the target a humiliating four times. Later, we had rapid fire on a better day and consequently, without our fingers wrapped in gloves, we did better.

One cold, rainy morning we hiked three miles out to Defendam Range to fire at air targets, only to find we'd gone to the wrong range. Another five miles of rainsoaked marching brought us to Banholtz Range where, between firing, we huddled around like wet chickens. The following day was just as wet and twice as cold, and found us on the transition range—firing at targets up a hill, targets that came up unexpectedly at varying distances. No one was the least interested in what was going on. We built a fire and huddled around it as long as we could, then dashed out, got our firing over with, and dashed back.

The worst came when it became our turn to go up into the pits behind the targets and operate them for the next group who came to fire. This might have been halfway cozy and warm had there not been a couple of feet of water in each pit. There was no standing up, or you'd have been hit, so it was splash and freeze and wonder if you'd ever get thawed out.

Happily, this was toward the end of March and turned out to be about the last of our cold days. The rain continued and nights were sometimes cool, but now our days of thawing turned into days of roasting. It was six of one, half dozen of the other, except that maybe there would be some relief from this damnable Alabama mud during the dry periods. At the end of this particular period of miserable weather the captain complimented us by saying we'd done a remarkable job under the circumstances. He went on to say that the following Friday, which was Good Friday, we would have a big beer party, being as how we'd be gone from the Company area on Easter—so our spirits were upped a little bit in that respect.

Easter Sunday would begin the eighth week of our cycle and would, for us, be the last of basic. But what a week. It would be putting to field use all the combat techniques we'd learned. These maneuvers involved a hike to a distant range, and six nights of battle and patrol. It was to take place at a gigantic training area beyond McClellan and the name of the place became a familiar word on every tongue from then on. Morrisville!

Meanwhile, training went on. I found the carbine much more to my style, primarily because the carbine weighs four pounds less than the M-1. And when you multiply that four pounds by the distances to various shooting ranges, there's quite a saving in ton-miles. And toward the end of firing it seemed the ranges they selected were increasingly farther away. Round trip hikes of twenty miles grew to be common. On one such distant range, the firing of the bazooka was demonstrated. A sheet held out in a frame behind the weapon illustrated the rocket backfire. When the sheet was ripped into shreds, we resolved never to stand behind a loaded bazooka. Then a couple officers fired a live round—the only one they had for us—and it turned out to be a dud.

Our first infiltration course was a snap. We got into a trench which contained muddy water that reached to the tops of our leggings and poured in to our shoes. After the initial shock our feet received, it wasn't so bad. At the proper signal, we went over the top, crawling and keeping low. Bullets whizzed above our heads at a height of about four feet, and explosions went off inside enclosures every few minutes. We rolled over a log, under barbed wire, then crawled along on the hard ground, occasionally encountering a puddle of mud, then more barbed wire and logs. Total distance: 87 yards. At the end, we clambered into a trench, fixed bayonets, and rushed out behind the machine guns, growling and savagely charging the enemy.

This was the best substitute for a mud bath that I knew offhand. My cartridge belt was twisted completely around and when I took a look at the rifle—dirty, muddy and scratched—I figured roughly on four hours of cleaning. At my speed.

We fired at landscape targets with .22s and for record with the carbine. One morning was spent with the .50 caliber M-3 machine gun, learning its operation and assembly. We were told it weighed 80 pounds empty and a hundred with ammo. After lifting some of the parts, a piece at a time, we believed them. The water-cooled .30 cal. machine gun came up for observation. We also studied the firing of tracers, and some of the trainees demonstrated for the rest of the company. Since the tracer shell contains thermite—a substance that reaches intense heat—we put out five or six forest fires during that phase of rifle training.

We had a couple night problems that lasted till 2400, and without compensating time in the morning. These were taken up with the operation of patrols, scouting, azimuth readings, and compass directions. Next day, we slept on our feet. That became a habit, though, in the last days before Morrisville. We were hurried around so much that rifle cleaning was left until the wee hours of candlelight or done in the latrine. Letter writing was also left till late at night, and when the dayroom was too crowded or when it was closed, everyone adjourned to the latrine and continued writing.

The result of this was a company of sleepyheads during training. A typical example was during an outdoor class on movement of regiments, when practically everyone fell asleep. It was such a farce that the captain made us all run around the hill once to wake up, but when we returned from that the nodding heads continued. Cadre and lieutenants circulated around tapping helmets, but eyelids couldn't keep from drooping.

And then, at length, we became combat infantrymen and the time arrived to put our knowledge into tactical field practice. The Battle of Morrisville was about to begin. We looked forward to it with anxiety and a little dread. It would be the toughest going we'd had yet, but by this time we were partially inured to hardships and perhaps it wouldn't be bad at all. That's what we thought. If only we'd known better.

At any rate, as April came to Alabama, we waited.

SEVEN
MORRISVILLE

Bob Hill and I got into Anniston April 5th, hitching a ride from the camp with a middle-aged dowager and a "cute little dog" named Victory who crawled all over our fresh clean uniforms.

Anniston, located about ten miles west of Ft. McClellan, was a thriving burg of 10,000 with an impressively beautiful residential section. But viewing the business district at night, as we did, gave us the impression it was neither large nor modern. Here we stocked up on supplies for the coming week: candy bars, chewing gum, cracker sandwiches, and other odds and ends. We had heard that a large part of the subsistence would be on C rations, so it was then up to us to plan ahead and see that we didn't go hungry.

Next day was a soft day and all of us had spring fever. That evening we dressed in our Sunday best and walked into the mess hall for our party to celebrate leaving Easter Sunday for Morrisville. As we passed, our plates were filled with fried chicken, dressing, and spaghetti. The tables were loaded with beer, pretzels, peanuts, candy bars, fruit cocktail, and cigarettes. We ate and drank heartily—I downed four Cokes—and afterwards simply relaxed, drinking, smoking, and eating peanuts and pretzels. The company talent was employed for entertainment: guitar, trumpet and singing, all in all as pleasant a two hours at chow as we could want. Beer went fast, and the captain announced that two more kegs were coming, which brought a cheer that raised the rafters a couple inches. Of course, we all knew what this really meant. But it was a nice way to be fattened up for the slaughter.

We had a night problem passing through a gap in enemy lines, then some squad tactics, field sanitation and camping, and on Saturday night before Easter the camp was a flurry of flying men preparing for the trek. Barracks bags, to be taken out by truck, were packed, clothes washed, packs rolled, everything being put in last minute readiness. Beds were crammed full so that we could see if we had everything, then unloaded into the barracks bag. I stuffed mine with overcoat, clothes, candy, gum, prunes, raisins, and other odds and ends. Next day, we cleaned up the huts and locked them, and fell out in full field equipment. At last, the time for leaving came.

The weather was just right as we marched away from civilization: cloudy, calm and cool. We walked at rout step (no particular cadence), chatting back and forth. We left the gates to McClellan and struck out along a country road. Spring had turned the landscape into a masterpiece of beauty. There seemed to be deep dry grass everywhere. Trees, vines, and bushes had leafed out in a luxurious green, mixed with pine trees to give a tantalizing aroma to the land. We passed along hedges made fragrant with honeysuckle, and alongside scrubby ravines. Occasionally through the forest that lined the road, we could make out farms nestled among the distant, rolling hills.

Evening came, and a cloudy twilight darkened the landscape. We passed a number of farm houses. The occupants lined up on broken porches to watch the passing troops. Behind the dilapidated barns and back houses were fields in which cows stood munching grass. We passed one farmhouse where the farmer had evidently amassed his family on the porch to watch us pass. I've forgotten what the count was, but there must have been twenty children.

We stopped for a break in front of another house, where the occupants were holding an Easter picnic, kids romping around and fighting. One of the women held up a beautifully colored egg and said: "Duz y'all wunt uh Eastuh egg?"

As the last traces of dusk disappeared over the hills, the night clouds grew ominous. At our second break it began to rain. By now, we were too exhausted to unroll the pack and dig out our raincoats. Besides, the rain wouldn't last long, we said. As we started the trek again, the sprinkles actually did stop. The temperature rose, the air continued sultry, and we became bathed in sweat.

But this was only the calm before the storm, and shortly afterwards the clouds burst full upon us. Rain lasted for the next ten miles. At first we ducked our heads against it, then gave up and slogged along, drenched to the core. I started to say something once to a buddy alongside, but only a sputter came out and I quit that. Every now and then, we would plunge into a puddle and splash mud shoulder high. By the time we finished the 13 miles to Morrisville, we were about as dilapidated as we could get.

We arrived at the bivouac site and had a fairly decent chow of lunch meat, cheese, cole slaw, potato salad, fruit cocktail, bread and coffee. After that, Bower and I put our tent halves together, pitched our tent, and hit the sack. Tomorrow would be the first of the big days.

The following morning began with a class on mines and booby traps. The company sat on stairstep seats in wooden stands with an open area in front, as though to watch a baseball game. After brief explanations by instructors, various types of explosive devices were detonated to give us an idea of the effect—anti-personnel mine, Bangalore torpedo, rifle, and fragmentation grenades. At the break we went down to examine the craters left by the explosives. The second hour, devoted to the study of high explosives, was conducted by a lieutenant whose technique and stage presence taught us more about explosives in that one hour than any other instructor had ever been able to accomplish.

For an introduction, he brought out a yellow half-pound can of TNT and set it on a table before him. He picked up a hammer, looked at the TNT, looked at us. Suddenly, we wanted some place to hide, some place far, far away, if this kind of thing were going to start. Really, if the fool wanted to blow his head off, we didn't wish to share in the project. But we were trapped, and though too young to die, we sat there and got the hell scared out of us when he brought the hammer down on the TNT with a resounding smash and shattered the explosive into about ten pieces. He pounded it some more so as to make a powder, then struck a match and threw it on the powder. My hair had already raised on my scalp. I could sense my classmates cringing on all sides.

Then he brought the burning explosive closer so that we could get a good look at it, as if we wanted to get any closer to it.

"Now you know," he said. "TNT can't be detonated without a blasting cap. However, don't breathe in those fumes from the burning powder. They're poisonous."

Under his guidance, two of the trembling trainees prepared a grenade out of a half-pound can of TNT, then stepped about ten feet back to set it off.

Before the fuse could be lit, the lieutenant said to the company: "How many of you fellows think this is a dummy?"

A couple raised their hands.

"How many of you fellows think this is a dummy?"

"Okay," he went on, "you boys who think this is a dummy come stand on the other side of that barricade."

The doubting Thomases turned white, shifted a bit and kept rooted to their seats. Then one of the volunteers lit the five-second fuse and threw the grenade over the log barricade.

We sat petrified, eyes wide, waiting, some of the troops with fingers in their ears.

After five seconds had surely passed, the lieutenant started walking back to the class, saying: "Of course, that was a dummy. But you can be sure that the enemy—."

BLAM!! From behind the barricade came a deafening blast, and a cloud of black smoke, flame and earth shot skyward. Two hundred men shot skyward at the same time.

Next: mines and booby traps. "You never can tell where a booby trap or mine might be," the instructor went on. "Why, there might be one over there..." He pointed his finger to a site close to the stands. **BLAM!!** Up went the ground in smoke and flame in a burst of TNT.

He went on. "Or maybe under that tree..." We cringed. **BLAM!!** A burst roared up under a very nearby pine.

"Or perhaps under those very stands on which you're sitting."

Whereupon everybody clenched his teeth and put his hands over his ears, drew his head down between his shoulders, tensely waiting to be blown sky high. But the lieutenant went on.

"No. Last time we blew one up under the stands, there were too many casualties and we had to make repairs. You can see where the wood has been chipped."

By this time, each of us had become a nervous wreck and got to the point where we didn't care what happened, though the instinct of self-preservation was still with us.

The lieutenant gently picked up a live British land mine and, holding it with reverence, said: "Now, always make sure you hold this package of sudden death still and move it very carefully, because it is one of our most powerful mines."

Then he threw it up in the stands.

While it was being passed from one to another in an effort to get rid of it and while the rest of us were getting ready to jump off the stands... **BLAM!!** A blast of TNT went off directly behind us, throwing smoke and dirt over the entire company. After the color returned to our faces, we roared with laughter for five minutes.

"Oh, him? Just came from explosives class."

We went into various areas for practical experience in mine work and were organized into teams that would work during the coming night in clearing a gap through a mine field. At 7:30 we moved out and fell into the stands near the mined field for a brief orientation. The first man out would be the pathfinder; he marked his path with a white tape, feeling around for mines and traps. Then the first wave of probers must follow, searching the ground with bayonets, locating other mines. A second wave of probers were followed by the lifters, whose duty it was to take out the mines and attend to any that had been booby trapped. I landed in a squad of lifters, and began to wonder if my insurance had been paid up.

Naturally, in a venture as dangerous as this, we were told to expect a number of casualties. The captain said that the only way we would know if anyone became a casualty when a mine went off, was to listen for a slight pop, which sounded when the percussion cap was allowed to drive home in the primer. So we took off and approached the field of doom.

There was dew in the grass and a nearly full moon crept up in the eastern sky. The pathfinder and probers went out and then it was our turn. We spread apart, just enough so that we could keep in sight of each other, and in sight of the man in the center, who followed the white tape. Mine locations were noted by the probers ahead and we began to dig. All around us, at relatively safe distances, TNT explosions were being set off, probably to get us in the mood. I began to dig gingerly, working a smooth circular hole all around the mine and looking for indications of booby trapping.

"Friend or enemy?"

Finally, the mine came out. I laid it on the ground and marked it. After lifting a couple more and not hearing the little pop that would mean I was a dead pigeon, I went along like a veteran mine-sweeper and began to pity the poor Japanese or Germans when I got over there. In a critique held after the problem, we discussed any difficulties that may have arisen or any obstacles we might have encountered.

To bed at 11:30, up at 4:45. The 1st and 2nd platoons went out on the combat course while the 3rd and 4th remained in the bivouac area to police up, clean rifles and sleep. In the distance we heard the muffled sounds of shots and explosions. That afternoon, we ran the combat course, and, naturally, it rained. A lieutenant who looked like Boris Karloff and acted like Frankenstein gave us an orientation on what we could expect to find. We received a dummy grenade and eight rounds of live ammunition. Then they bade us farewell.

We walked slowly through the woods, alert, tense, ready. We slid under barbed wire in deep mud, crawled over a wall, dived into a pit to bayonet a dummy, and then jumped into a trench of knee-deep water. I didn't hesitate jumping, but on the way in it occurred to me that the water might be ten feet deep.

Each of us went through the combat course alone. After floundering in the trench, I crawled out, only to be confronted by a dummy to charge, then a machine gun nest at which to pitch the grenade and charge. I ran up to the nest and jumped in. Here again, three feet of water, but I had become so soaked that the geyser of water shooting up didn't mean anything. Afterwards, dripping, we all gathered for the customary critique.

When the week was half through, I began to think we'd overestimated the rigors of Morrisville. It had been a snap thus far, and surely it wouldn't get any worse. But then, it turned off cold, and crawling over an icy, wet ground didn't help matters. Just the same, while I was in reserve once on a night problem, I flopped on the ground and went to sleep for a couple minutes and never got pneumonia from it. No matter how cold it got outside, the tent and blankets, what little time we spent in them, were warm as toast—until the temperature *really* fell. We had squad tactics in taking a machine gun nest, then in taking a hill in "Squad and Platoon in the Attack." Then Morrisville took on big scale dimensions. Five thousand soldiers stood on the battle field and watched a bomber and fighter air show above.

On Wednesday night, the cold grew intense and crept into our tents so that no one got much sleep. Thursday was spent in more squad tactics and intermittently listening to instruction and sleeping wherever the slightest chance occurred. We came to the point where during a two-minute lull we got 1½ minutes sleep. After enough of these, we began to wake up and regain some spirit for the problem that night.

The entire company marched out at rout march, five yards apart on opposite sides of the road. There was no talking whatsoever. Flank guards were put out from time to time so that an approaching enemy would tangle with them and warn the main body of troops. This turned out to be a tougher job than it sounded. When my time came to go out at a certain turn in the road, I crept confidently off the road and at once sank to the knees in water and mud. I floundered around just enough to decide on rejoining the march to try it again later.

A few yards farther on, my marching mate—a fellow by the name of Campbell, from Denver—broke off with me and we found ourselves in a huge thicket of thorn bushes, probably blackberry, on a steep hillside in pitch blackness. I lost my footing and began to go down. I stumbled, fell and started rolling, my rifle coming down behind me of its own accord. When I picked myself up, I drew a couple of thorns out of my face and hands, found Campbell, and we continued a few yards farther. In due course, we decided that the way was almost impassable here without being torn to shreds, so we staggered back up to the road, hurried along, and rejoined the company. What a couple sad sack flank guards we would have been in such terrain! The more I thought about it, contact with the enemy might have been preferable to contact with Alabama blackberry.

We arrived at our destination, a small grass-covered draw surrounded by tall pines and heavy underbrush. A couple volunteers went out on a mission. The rest of us sat there and had nothing to do but shiver while we waited. The lieutenant decided he'd have to start something to keep us awake and the only field expedient he had handy was the sky above, which happened for once to be clear. As the only amateur astronomer in the group, I was trapped into giving a star-gazing lecture. I had credentials. I'd been a frequent visitor to the Lowell Observatory, on a hill above Flagstaff, Arizona. In fact, one night, the celebrated astronomer, Clyde Tombaugh, discoverer of the planet Pluto, opened up the observatory's 27-inch refracting telescope just for my buddy and me. That was a thrill few teenagers could boast of. The following week, Tombaugh sent me a post card picture of the Observatory, signed by him and by Earl Slipher, another eminent astronomer on the staff. I had guarded that like others guard baseball cards.

In any event, becoming an impromptu instructor kept me awake, too. I pointed out the Big Dipper, Orion, and other constellations, showed how to find the North Star, then took questions.

"By the way," said one of the instructors, "can you point out the Little Dipper for me?" After I pointed it out, he said, "Well, I'll be damned! I've waited 32 years to find that..."

On Friday morning, time was devoted to preparation for movement of the bivouac site and for an inspection. We were told it was only going to be a pack inspection, and I consequently didn't clean my rifle. You'd have thought that by now I'd have been on to this little trick of theirs, but I wasn't. Rifles were inspected. Curses skyrocketed when it came my turn at inspection arms.

That evening we left at rout march. I was placed as connecting file between the point and the first platoon. This time, the pack bore down like never before. In addition to all the full field equipment, there were six cans of C rations wrapped in an extra blanket over the top of the regular pack. If you don't think that was heavy, you might walk in and lift the kitchen stove to get the idea. Besides all the weight, it was raining and everything became watersoaked, adding more pounds.

This lasted for six miles. Who the hell said Morrisville was a cinch?

Along the way, I saw lights in the distance, lights that blinked at irregular intervals and which I took to be flashlight signals. I was about to give the alarm of an enemy approaching when it occurred to me what the lights were. Fire flies. They flitted among the dripping leaves, their lights flickering, on, off, on, off—tracing the stygian dark with their tiny glow.

Upon arrival at the bivouac site, it began to rain in earnest and we pitched our tents in a downpour. No one got much sleep that night. I slept on a cold, wet raincoat under a cold, wet blanket. We arose early and marched a mile to the RJ maneuver area, a mile-square, sloping valley with perfect battle training conditions and an observation post with stands overlooking the panorama. In the center, Hill B; to the right at about 1,200 yards, Hill X; to the left, Hill C. We ran it in the morning and it turned out to be the climax of our basic career.

I was placed as the squad leader of the 3rd squad, 4th platoon, whose object it was to form a skirmish line on Hill X, and then take Hill C. After we'd finished a grueling run of the problem, we were told that that was practice. So we did the thing again just after noon, this time with all the trimmings.

The terrain was decidedly rough. Our route lay through grass, then thick jungle, with thorn bushes and swamps. Then up a steep hill. There was a creek you could cross if you luckily found a log spanning it, otherwise you did as we did, jumped in, sank to the waist and waded across. Most of the progress was in double time, assault fire, and uphill to the other rim of the valley.

"You sure got to hand it to these artillery boys!"

Heavy machine gun and artillery fire exploded all around us. Artillery pieces were located back beyond the stands, hidden in the forest. But the shells from them certainly didn't hide. I hoped they knew just where they were firing. When one shell hit too close, I thought about taking off my shorts and waving a white flag of surrender.

They were firing on Hill B, and we kept our respectable distance from that area. Concussion swept across the valley. Smoke bombs went off, sending up a huge screen in front of us. Then came a climactic one-minute bombardment concentrated on the hill. When that stopped, we rushed up the hill firing. This same process was repeated at Hill C and by that time we were ready to drop dead.

After our conquest, we climbed to the stands above the valley and watched another group go through a repeat of the entire exercise, machine guns, artillery and all. By then, we were experts, of course, and able to pick out all the mistakes of the others. But we got so overwhelmed by the spectacle that we forgot about mistakes.

Tired as we were, we wondered how the Infantry companies stood up under two weeks of this. When the last big show of the day ended, we marched back from the RJ area, packed our gear, and prepared for the march to McClellan.

The stiff workout of the day—running, stumbling, climbing, crawling, and plunging forward in assault—had taken its toll on a number of men in the company and they were limping badly even before beginning the muddy nineteen mile trek home. A few men, so exhausted they couldn't walk another step, were picked up by a truck and carried home. But for the rest of us rugged, hardened veterans, still able to stand on two feet, there was nothing left to do but march, march, march.

We packed, ate, policed the campsite, and at 1900, in the humid gloom of evening, set out on the road. After 45 minutes of walking, we got a 15-minute break. I didn't even think about sitting down, for the torture of getting up would have been too great. Then we went on. Nobody said anything. Nobody had that much extra energy. We slogged along in the pitch blackness, aware only of the presence of the figure straggling alongside. As a squad leader, I walked in front, with no one to follow.

You may not believe this, and I wouldn't, either, if I hadn't experienced it, but I actually *slept* while marching. Much of the way was walked with eyes nearly closed, as if about to go to sleep. I opened my eyes periodically, just to see if the dim outline of the road ahead were still straight. During several periods of five and six seconds I actually did fall asleep, lulled by the monotony of putting one foot in front of another, and by the soft sound of footsteps on the damp earth. Then I waked with a start. Thus, I became convinced that there was something to the saying, "out on your feet."

For the first ten miles, it was sultry and hot. Sweat poured off in buckets, soaking us to the skin. Fire flies jerked along through the forest on the sides of the road, but we didn't pay much attention. Eventually a kind of relieving numbness set in and, during the breaks, we flopped down flat on the road and slept. We were so tired that, on rising, we couldn't feel anything any more. Occasionally, the darkness was pierced by headlights, and a truck passed us loaded with stragglers who had literally dropped by the wayside farther back and couldn't walk another step.

At 10:30, it began to rain, but we scarcely noticed. I put my field jacket around my neck and made that area the only dry place on me. Soon the road became a sloping muddy path through the woods, and we slipped and slid with every step. Not bothering where we put our feet, we slopped and splashed along and were soon covered with mud. The packs, already heavy with distance, now grew soaked with water and our shoulders began to come apart.

Once somebody heard a groan and flashed a light off the road. A GI lay there near a culvert, sunk to the hips in mud, waiting for the next truck to come along. He looked up and seemed to wave us a weak greeting, but he did it only with his eyes, for he was apparently too tired to move a single muscle.

At length, the mist ahead began to brighten with the lights of McClellan. As we approached, the rain increased to full intensity, and an electrical storm hit with maximum force. Thunder boomed and echoed incessantly in a deafening roar.

The lights of North Gate came into view. Half of the platoon was limping and none of us could do more than squint in the downpour, but we all sent up a rousing cheer of joy as we stepped onto the pavement after nineteen miles of muck. We yelled again as we entered the gate, but the sound was lost in a clap of thunder.

Rain fell in buckets. Lightning zipped everywhere, giving the impression of more light than dark; continuous blinding flashes that lit the entire camp, as if to welcome us back into its protective folds.

We arrived in the company area and stopped in formation on the street in front of Hut 9. The sergeant held us there a moment, drinking in the sight of some of the most slumped and bedraggled figures ever illuminated by brilliant lightning bolts. Not all the platoon was there, but by far most of us had made it back. We looked like broken derelicts on a shoal at sea.

Then he said, in a low voice between thunderclaps, "Dismissed."

The yells that followed shook the camp. We plunged into the hut, now a chunk of heaven for our battered souls. Dropping our gear in the nearest empty place on the floor, we slithered out of the raincoats and shot across the street to shower off the mud. The mess hall opened up to serve coffee and doughnuts, and a few of us stumbled down to eat. When I finished, I decided that I was tired, so I crawled back to the hut, and dropped dead in my bunk. I am prepared to say, without a shadow of a doubt, that all through the rest of the night, and late into the following morning, not a muscle moved. Not once.

EIGHT
RADIO

"'Twas the morn after Morrisville, and all through the hut, not a creature was stirring, not even Sut..."

I pried my eyelids open and started to look around. That was a mistake. I tried to turn over in bed. That, too, was a mistake. After half an hour I loosened up enough sore muscles to prop myself up and look around the hut. And the only expression that came to mind was: "What a mess!"

I was referring not only to the rifles and gear heaped up everywhere, but to the sad sacks still sawing logs in the bunks. Everywhere, the dispelling of fatigue was taking place. Every conceivable bedside contortion was on display—mouths open, feet draped off the bed, faces plunged under pillows, covers on the floor—not what the Army usually approved of, but expected under the circumstances. Here was nothing but total relaxation. Outside, the warm Sunday sun was shining, the birds singing.

Soon the others began to stir. I propped myself up on the pillow to witness what was about to take place. Evan Marsh waked first and started to move a leg. The groan that followed waked up everyone else in the hut. Then, as the others pulled themselves up in bed, the whines, grunts and groans became general. There were squalls of surprise and shock over a paralyzed leg or arm.

I burst into laughter and was hit by several pillows. Then I got up. I almost fell out of bed. I put one leg out and, since there was no muscle power to hold it up, it fell to the floor with a crash, and I almost went with it. I examined briefly for broken bones, then rubbed some sore spots and investigated my feet. I was lucky—only two blisters. Others were not so well off, generally much worse, including one poor chap whose feet were a mass of blisters. We got to talking about the hike, and some said that along the way they'd gotten to hurting so much they almost cried. But all came to the same conclusion that the full day's work, then the march home to top it off, was no cinch in any man's language. I fully agreed, but played the wise guy and said: "Ah, you sissies just can't take it, that's all. Let me take you down the trail into the Grand Canyon some time." I got hit by another barrage of pillows.

Clothes and equipment were washed by men who could barely stoop over the laundry tub. They were laid out in the street to dry, and we had to pick our way carefully going to and from the latrine. Everyone limped, in some fashion or other, and you'd have thought it a rheumatism epidemic in an old soldier's home. That's the way men are trained to fight.

The next two days and nights were devoted to review and we bivouacked in a little valley outside McClellan. The area of review was atop a steep hill and climbing this brought back some of the sting of the as yet unhealed Morrisville wounds. We spent an hour at each of sixteen stations during the two-day period and covered first aid, bayonet, machine gun, anti-tank, aircraft, chemical warfare, hand grenades, camouflage, technique of fire, squad tactics, and others. With this, basic training came officially to an end. Somehow, none of us was at all sad about it.

The time came to turn in our rifles, an utterly delightful task. There were grim smiles and leering countenances as we walked up to the supply desk and shoved the weapon—our best friend—across the counter. Happy, relieved faces turned away after checking in everything. Peace at last, we said, for the rest of the cycle. No guns, no cleaning, no inspections, no gigs—nothing but rest and relaxation.

That was on April 18th. On April 26th, we got our rifles back again.

Whatever had been the hardships of basic, we couldn't deny that we'd learned a lot. Of course, we may have been just glorified Boy Scouts and probably deserved several merit badges for our trouble, but it was good training and had a far different underlying purpose. One thing it had done for me showed up one day on the scales: I had gained 20 pounds since MacArthur. And brother, it wasn't all jam either.

At last, technical training began and we as a platoon of solid radio became an integral unit. The cycle settled down to an hour of physical training and seven hours of classes each day.

By far most of the time was occupied with learning Morse code, a collection of long and short tones that represent the alphabet and can be readily transmitted over radio networks. The Radio School was a large frame building with long tables running lengthwise inside. Each position at the tables was walled in, so that each writing place amounted to a semi-private cubby-hole. Each such nook contained a pair of headphones and a code sending key. At the end of each table lay pencils and stacks of paper pads for practice in translating code transmissions into longhand. At the west end of the building was the control room, so wired with the main room that any instructor might monitor any individual position at the tables.

Shorty and the "Tarpaper Palace," Fort McClellan Radio School, 1944

One side of the control room was a mass of tape-playing devices from which lessons and code practice were piped to the tables. Entire tables, or parts thereof, were electrically partitioned to accommodate the number of men currently on the same lesson.

We started out with preliminary instruction and then went to Lesson l. Here was the introduction to the alphabet as well as to the standard phonetic language (Able *di-dah*, Baker *dah-di-di-dit*, Charlie *dah-di-dah-dit*, Dog *dah-di-dit*, Easy *dit*, Fox *di-di-dah-dit*, George *dah-dah-dit*, Item *di-dit*, Jig *di-dah-dah-dah*, etc.). First learned were the letters M-H-J-R-U-F-G, so arranged as to correspond with those letters hit on the typewriter by the index fingers, and to facilitate the taking down of code signals when using a typewriter. We, however, got all our training the manual way—with paper and pencil. Then the other letters and the numbers followed and were completed in five lessons.

During these first hours at the code table, we sat and listened to slow transmissions, such as: *Voice: Charlie—dah-di-dah-dit*, dah meaning a long dash or tone and the di or dit meaning a short dash or tone; then a cw signal, i.e., one transmitted as a signal on a carrier wave, five times. That engraved the code form of the alphabetic letter onto our brains. As soon as we got the hang of the entire alphabet, and started associating each letter with the correct arrangements of dits and dahs, we got into speed work. This built up capability to absorb a stream of dits and dahs and write the corresponding letters out in longhand. As the days and weeks passed, we gained the ability to recognize a letter instantaneously, without thinking. Thus, when we heard the signals *dah di dah dah, dah dah dah, di di dah* speeding over the wire line, we wrote YOU as fast as we could. During training, however, we received random groups of five letters, for example: WM3CP 2CEOB J972S RTPPN U4ML0. With this, code-taking became competitive, and most of us strived like mad to receive faster transmissions, measured in code groups per minute, to see if we could get ahead of our buddies.

Tests were sent by skilled operators from the control room, and when each of us felt we'd passed one by getting all the letters (here sent at random) correct, we shot our papers in the air to be collected by a noncom (noncommissioned officer). Then we chewed our nails to see whether or not we'd passed. After completing the alphabet in the initial five lessons, we started at six wpm (words per minute, meaning code groups per minute) and strived to increase our speed. My record of successful tests went this way:

Lesson 1	**April 21**
Lesson 2	**April 22**
Lessons 3-4-5	**April 24**
6 wpm	**April 26**
8 wpm	**April 28**
10 wpm	**April 29**
12 wpm	**May 5**
14 wpm	**May 8**
16 wpm	**June 13**

That I was stuck on 16 wpm so long worried me at first, and I couldn't explain it. We had learned a system of writing each letter of the alphabet in the fastest possible way. But it took a while to remember that A and F, for example, were each to be written in two strokes, not three. We had to write legibly. If we didn't, if we got one letter wrong, or even slightly fouled up, we failed that test.

I kept plugging away. After all, I told myself, I had the duration plus six months to pass it.

Taking code practice four, five and six hours a day brought to us a saying unique in the organizational lexicon. When we saw someone doing anything a bit unusual, we classified him as "dit-happy," and kept him under observation. Moreover, the sergeant said, we could not expect a lessening of the condition, for we were slated for 200 hours of code practice.

Our training, however, was far from straight code. Radio transmission procedures and the care and operation of radio sets came in for a great share of the time. And just as the other platoons of the company must take some code and radio practice, so did we undergo lessons in message center operations, telegraph and wire techniques. We learned visual signaling by cloth, flags, and lights. Radio net operation took as much time overall as code. At least it was just as important, for learning Morse was one thing—putting it to use over the air was something else.

Though summer was on its way, we could still count on half of each week to be filled with that liquid Alabama sunshine. Water wings weren't regulation issue, or we could have put them to use. And then, at last, came the days of heat and we changed to summer khakis. It was a damp heat, of course, and sweat poured from us, even after the slightest exertion.

The route from Hut 9 to the Radio School lay uphill, so we had to fan ourselves each time before starting code practice. Wearing earphones formed a pool of perspiration around our ears and this had to be swabbed every so often.

Company details continued—guard, latrine, and KP. While I was able to act as corporal of the guard and thus not have to stand guard, being a squad leader opened the way to the additional chore of CQ at Battalion Headquarters, that is, in charge of headquarters during off hours. But this had its advantages, for the Battalion offices had several typewriters and I often postponed letter writing until those nights and pounded away into the wee hours.

Inspections continued to plague us, and once in a while we had some night training, but on the whole we basked in the luxury of having a great deal of free time, as compared to basic. I experimented to see whether going to bed early would make me the wiser and sharper during classes, but after falling asleep during a number of classes, I accepted the proof and stayed up late from then on. You can get more loafing done that way.

In fact, I actually went down to Newcomen Library on the Post and checked out a book. An even bigger surprise was that I got to *read* it, something that could never have been accomplished during those first seven weeks.

We visited our friend in Company A-9 as he entered his 10th week of basic, and it seemed the routine had gone on long enough for all of his associates to become bitter. When we hinted at what we were doing, he practically fell over. We tried to play it down by talking of our Morrisville hardships, but that didn't help. His company was slated for two weeks of Morrisville. That, brother, was the end.

We used Signal Corps radio sets for instruction and field training, and began to communicate with each other over short distances. We were introduced to sets SCR 195, 511, 536 (handie-talkie) and 300 (the walkie-talkie, an FM-AM set). We thoroughly enjoyed circulating around the Post and operating our simple networks. Gattis and I were together once with phone communication—I climbed Regimental Hill while he traveled to North Gate and other positions. We sent messages by code and practiced voice procedures. It brought to mind that first day in the C-15 area when I was piling rocks and watching advanced radiomen do this same thing, my face green with envy. I knew now. It was as nice as it had looked.

We perched on elevated points in the field, lounged on the soft Alabama grass, listened to the radio and, once in a while, transmitted "Sugar, this is Sugar One. Message for you. Over."

Soft life back in the company area came to an abrupt halt when we began preparations for the one big inspection to end them all. This was General's Inspection, in which not only we, our huts and our company were to be examined, but the progress we had made in technical work as well. *Everything!*

So, morning and evening, company inspections became common, with one purpose in mind... to slap those who got gigged into a detail of cleaning the company area. This meant either clean and polish your own equipment, or you'd be doing both yours and the company's. Such details brought special attention to the latrine, day room, mess hall, and orderly room. They were cleaned and scrubbed to the last square inch, inside and out.

Our huts, of course, got the same attention, a job that fell on the occupants of each. We took the stove down for the summer and cleaned and mopped and scrubbed and polished time and time again. The roof was swept, the screens washed, the windows polished like astronomical mirrors. The hut was constructed on posts; the posts were scrubbed and the ground beneath the hut policed to the last scrap. Nothing was humdrum during this period. Every evening during the week before inspection was one of hustle and bustle. We turned in our field jackets to be cleaned and pressed and got brand new bayonets and cartridge belts. Add to it memorization assignments from radio procedure and night review periods. Sleep grew less and less, and eventually we were not permitted to leave the company area at all. We became the shiniest soldiers with the shiniest company area in the Army, and on May 13 the inspection came.

It lasted all day. We fell out with clean pressed khakis, brand new leggings and glossy, mirror-shined shoes. Rifles, it seemed, were cleaner than ever before; perhaps, in my case, because I had learned not to run, stumble and plunge them into the sand. After we had arrived at the parade ground and lined ourselves up, the general came. He was a brigadier, impressive to us because we'd never seen *any* stars so close. With him were his staff members: colonel, lieutenant colonel, major, captain, and lieutenant. We'd never seen so much *brass* in one place, ever.

The general passed along, asking questions and requesting demonstrations every so often. My being up front as a squad leader, and being six feet five in height, almost guaranteed that I would be singled out. The general came toward me and I went white. He stopped in front of me and asked softly that I execute a right-shoulder arms. I complied, almost dislocating my shoulder. "Very good," he said, and I rose up into the clouds. At noon, I looked down my nose at the rest of the men. I mean, after all, how many rank and file get to hobnob with so many officers?

General's Inspection of our technical training turned out well. The Radio Section made an all-time record score of 98 out of 100, and then we all looked down our noses at the rookies of the other platoons.

After the inspection, we went back to the soft life again. As code practice wore into its advanced stages and we sat on the higher speeds, the continuous monotone had a lullaby effect, highly inducive to sleep. Instructors in the code room circulated among us, looking for nodding heads. They had devised a series of diabolical methods to wake someone dozing. They would pick up a stack of writing pads, come to the cubbyhole, poise the pack about five feet above the table, and drop it.

The resulting crash sounded like a 60mm howitzer going off in the victim's ear, and impressed the wrongdoer as to the evils of falling sleep. We couldn't believe our eyes one day when a stack of pads was dropped on one poor soul's table and he didn't wake up. The pads were dropped again, with an even louder crash, and still he didn't stir. Then the instructor dropped the whole stack on top of his head. That did it.

It was about this time in the cycle that rumors began to fly thick and heavy in regard to furloughs and other destinations upon completion of the current cycle. The most widely discussed story concerned advanced training. It became general, though unofficial, knowledge that if you were at the top in code speed, had made good grades in classroom instruction, and had other merits that classified you as a competent radio operator with "possibilities," you would be sent to advanced radio training, probably at Fort Benning, Georgia. The next logical step would be a boost out of the Infantry and into the Signal Corps. But then, we were still green to the tortures of rumor-believing and liked to dream anyway.

"Come now, Throckmorton, I think we can find other ways to keep the men awake."

Still, I began to get one of those feelings—a kind of gnawing desire—that such a dream might come true. That I might just accidentally get to go on and learn more radio. A number of us began to get the same intense feeling. To make it worse we heard this: *B-15 finished their cycle while we were in the middle of ours, and 20 B-15 Radio men were sent to Benning.*

A greater majority of us would gladly postpone our furlough four months, just to go on to school. Gladly.

On the other hand, rumor had it, those who didn't go on to advanced training and wanted to go home must show the $150 in cash they planned to use for the round trip. Our first sergeant, the top of the non-coms, verified the fact that the Army wasn't paying the way home any more—in many cases—for its men on furlough. So furlough or training, we stayed up in the air about the whole matter.

All the guys in the hut deserted me in writing to the girl in Iowa, traitors that they were. I was still her "dream man," though, and one day she sent a recent picture of herself.

Aubie Darr, the playboy, took one look and said, "Boy, is she *stacked!*"

Good cook, too. She sent a box of cookies, and a box of 15 big candy bars. Naturally, when that happened, the whole hut claimed to have had a stake in getting me introduced to her. I said, "Nothing doing!" But I did relent, and everybody pitched in. After that, though, our correspondence waned, for school was getting tougher.

Summer came in full force and we began eating salt tablets to restore the sodium chloride we lost through perspiration during exercises. Before the Post pools opened, some of us found a deserted training lake back of camp and went there often to cool off.

Technical training continued. We practiced splicing telegraph wire and the installation and use of field telephones. We had code keying practice and developed our sending fists. When radio procedure practice ended in the classroom, we took it to table nets (several operators communicating with each other) in the code room and then out with our radio sets into the field. We learned the SCR 284, a portable set with attached generator, whose total weight exceeded 125 pounds ground installation, 200 pounds mobile. A box with legs, it held a 25-foot antenna, mounted by piecing together eight mast sections. We set up nets around the school, practiced mobile operations, and rode around the fort in radio trucks, operating nets.

The transmitting rules were laid down to us as the basis for operation, and we were never to swerve from them:

1. Listen in before transmitting to avoid causing interference.
2. Make only minimum transmissions necessary to maintain the net and to clear traffic.
3. Send all call signs clearly and accurately.

4. Transmit code at a speed no faster than that of the slowest operator.
5. Reply promptly to all transmissions requiring an answer.
6. Conform strictly to prescribed radio procedure and regulations for maintaining signal security.

Then came Q signals, radio shorthand to convey the most commonly transmitted messages or instructions. Example: QNW, shut down your station. Intermittently, code kept humming in our ears. Our platoon leader, a warrant officer, told us one day after class about the effect of all this steady code on an individual. With glaring eyes and quivering lips, he said: "It has no effect on you—much. Why, *I've* been around code for six years, and look at me!"

That was the trouble. We *were* looking at him. We burst out laughing.

"But," he continued, "there *have* been severe cases, such as the one where a fellow waked one night grabbing for paper and pencil to copy some code he thought he was hearing. But it won't affect you... heh, heh, heh..."

Next day, believe it or not, one of the guys came into the hut at noon and took off his leggings to start changing clothes for retreat, normally held at 5 p.m. It wasn't a gag, either. All of us did something or other at some time to convince ourselves we were not immune from becoming dit-happy.

One morning, we'd been out for three hours with the 284s and then settled comfortably in the code room for code practice when the captain came in and announced that 30 men from the radio platoon would go to Benning for advanced training. He asked how many would like to go and got a unanimous show of 39 hands. Then he said that the list would be chosen within the next two days, but probably final approval wouldn't come until the 17th week of training, or until the cycle was almost over. As to Benning itself, the captain had this to say:

"It's a three-month vacation. You get up at seven, eat if you want to, go to school from eight to five, have swell barracks, week-ends off, no KP or any details, no inspections or rifles..."

That was enough. By now we were panting and whooping too loudly for him to continue.

We spent one of the hottest afternoons at McClellan this way: The company was called out to the parade ground, where we took off our shirts and undershirts and struck off across country double time. This lasted ten minutes, or long enough for us to work up a lather of perspiration. Then we slowed down for a mile-and-a-half walk, all in stifling heat, enough to make our clothes so wet we were dripping. Then, suddenly, the captain stopped in the hardest, dirtiest, and dustiest section of land he could find and pulled that old favorite of all officers: "On your back! On your belly! On your feet! On your knees!"

We created a major dust storm that combined with sweat to form a thick coat of mud all over us. Then pushups. By the time we trudged back to our clothes, we gave the indication of just having pulled out of a mudbath. We were covered from head to foot. Mercifully, the captain didn't make us put our shirts on over all that mud. Instead, he took us to the Battalion latrine. We raced inside and got under the showers, clothes and all. I don't think I ever enjoyed a shower so much. When the latrine orderly on duty saw us coming, he almost fainted, but we revived him by saying that some of us would remain behind to help clean up. Thus it didn't matter whether there was rain or shine, the Alabama summer got us wet in one way or another.

Sometimes we lounged around in the cool of a cloudy day, sitting in the grass sending messages over field nets. From time to time, we saw marching columns of rifle companies, weary and haggard, and fell in with the old saying: "Brother, we never had it so good!" Nevertheless, we were learning a new trade.

During the 14th week of training, the battalion was rearranged so that each company would be an integral unit. Instead of having the four separate platoons in one company, they reorganized so as to make all four platoons identical. Thus C-15 became a wire company, and we were transferred to A-15, an all radio company. Of course, this meant that the various platoons would be in different cycles of their training, but that, we supposed, would be ironed out. We didn't worry about it because we were the veterans of the new outfit. One platoon was in the 3rd week, one in the 7th, one in the 11th, and ours in the 14th.

Since C-15 had gone so far together, we held one last party before we split up. They managed to get the North Gymnasium for it and planned on a bang-up time. Several of us went to the movies first and arrived at the party about 2230. After looking in, we wished we'd seen the second feature. There were some WACs and the dancing was good, but that was all. Beer was stinking all over the floor, and a large percentage of men around the walls were flopped over drunk. We watched a while, drank five Cokes, then became overpowered with the stench and left. In spite of this, we remembered C-15 as a great company, considering how it had survived Morrisville and General's Inspection.

The 14th week brought a finish to technical instruction. We climbed to the top in code practice, finished visual signaling, and intensified field nets and radio operation. Some problems lasted all day and some into the night, so that sack lunches had to be brought out to us. Eventually, we were ripe for the plucking, and once again we left civilization and communed with nature—for the 15th and 16th weeks were to be on steady communications maneuvers in the field. Way in the field.

So in the last remaining days before leaving civilization, we concentrated on the luxuries we would miss. We consumed ice cream pint after pint. We saw all the shows. We spent hours beneath the bright, cool blue waters of the north swimming pool, basked in the sun, and generally relaxed. We enjoyed it to the fullest. And we were glad we did. For suddenly, all the pleasantries of civilized life came to a screeching halt.

Shorty with Radio Set SCR-284 in Talledega National Forest, Alabama

NINE
TALLEDEGA

Maneuvers lasted from May 29th to June 10th. Whatever complaints may be made about them, at least the whole thing started out nice. We rode in trucks to the initial bivouac area, and enjoyed all 25 miles of the ride, thinking what it might have been if we'd hiked it. The problems were held in the Choccolocco Camp area of the Talledega National Forest, in that romantic backwoods country where the Blue Ridge meets the Appalachian foothills.

The first three days we spent going out in the vicinity of the bivouac area and maintaining field nets. It gave us a chance to get acquainted with the scenery. And the Alabama forest was beautiful in spring. Everywhere there was foliage, shrubbery, tall grass and wild flowers. There were birds and butterflies—generally a pleasing atmosphere to radiomen continuously listening to the monotone of dits and dahs. The trees were predominantly oaks and pines with a sub-community of maples. Saplings graced the shade and shadows. The terrain was far from level, yet neither mountainous nor hilly. There was just enough combination of rises and ravines and underbrush to discourage prolonged nature ramblings. For the most part, the sky remained a bright blue, with intermittent clouds floating lazily and a light cooling breeze blowing through the leaves with a cozy rustle. Perfect conditions for a summer outing.

We were not, however, on a summer outing. Business was strict. One word governed our movements and that one word we heard continually: TACTICAL. Eating, sleeping and all the work and moving we did tactically, that is, without revealing our whereabouts to the enemy. We camouflaged everything: tents, helmets, radio sets, mess tent, trucks, foxholes—yes, even the slit trenches. Concealment was a byword; during the next two weeks, reconnaissance planes would fly over the forest making photographs. Woe be unto us if anything showed up on those pictures.

So we set up our radios, during the first three days, above ground, with the antenna piercing the lacework of foliage above. For a writing desk, we used either our knee or the thick ground cover of pine needles, leaves and grass. The four operators assigned to each radio set worked in rotation—half-hour turns at the set, the generator, the log book, and asleep under the nearest tree. Rotation allowed spare time for shooting the bull and for writing. But at other times we kept busy. While being a squad leader had exempted me from KP and latrine duty during the last weeks of technical training, things changed now and there was little rest for the wicked. In the field here, it involved additional duties that kept us running. The sergeant called us in for night conferences to give us the lay of the land and how to adjust the training procedure. We circulated to our crews then and checked foxholes and other details and spread the G-2 (scuttlebutt) as to what was going on.

Digging prone shelters for use in case of air attack turned out to be one of the most difficult phases of tactical bivouac. Digging among rocks and roots with tiny shovels was bad enough, but the dimensions had to be six feet long, two feet wide and four feet deep. Of course, I had to remove an additional cubic yard of dirt and roots and rocks to get my trench long enough. The reason for such dimensions we surmised to be that when you dived for the hole and your aim was bad, you had a little leeway to bounce once and fall in. I'm glad we didn't have to try it.

"No, you cannot have a jackhammmer. Get to work!"

Then came the miniature problem that lasted from Thursday noon until Saturday. It was a CPX problem in which battalions were set up and moved along under simulated battle conditions. There was no actual firing, but things were hot over the earphones. That first night, perhaps the worst, messages began to pile up and every station operator in the net tried to get his message in edgewise. Being net control station kept us on our toes the whole two days. I think I fudged once and stole five hours sleep. The three platoons—message center, radio, and wire—worked in full cooperation in getting orders through by whatever means.

The intelligence platoon played the role of the enemy. Its members sneaked up, and if they succeeded in finding your station, they marked red X's on equipment, cut hammock supports, and committed other deviltry. If they could get away with it, without getting captured by us, they were supposed to be good and we sloppy. Any radio operator who fell out of his hammock because of cut strings vowed to castrate the next infiltrator he caught.

Regimental headquarters was set up as a controlling point and the 1st, 2nd, and 3rd battalions kept moving ahead, leapfrog fashion, contacting the enemy. Messages went first to the message center for conversion into code by the M-9 converter, a coding and deciphering device which was at the time a strictly confidential piece of equipment. Then a messenger stumbled through the underbrush to get the messages to us for final relay.

The officers in charge of the mythical battalions kept the lines of communications busy. In our little hidden radio stations, papers flew all over the place, while telegraph, telephone and messenger service went full blast. At night, the two radio operators on duty would huddle beneath a canvas as long as they could, breathing smoke from the candle flame used for tactical lighting. Each would frequently come out, gulp fresh air in great gasps, then plunge back under again.

During this miniature problem, we picked up and moved four times. This involved cutting out of the net, closing down our station, packing up the radio gear, and lugging all the equipment on foot (usually at night) through the Alabama jungle. The SCR-284 radio set, in an olive drab metal box roughly two feet cubed, was no feather. I remember stumbling, turning around, and slamming it against tree limbs more than once. It sounded like a kettledrum in the profound silence, or a wrought-iron stove banging end over end down the hillside. My buddies whispered from the shadows: "*Tactical*, you bastard! We'll *all* get caught!"

Getting there was only half the problem. On arrival, I tried to let the radio set down gently, but it inevitably hit an unseen rock among the pine needles, and reverberated with another boom. Then we had to dig in. Silently.

If it has not as yet occurred to you who dug those radio pits, who sweated and toiled over pick and shovel to sink the radio set and its operators below ground level, then let me set this business straight. We did, brother, and we did it each time.

"Roger One, this is Roger Six. Over."

The customary prone shelter had to be dug first, and after that, we spelled each other in gouging out a hole that was sufficiently large to accommodate the set, generator, and two operators.

These pits became rather cozy havens of seclusion, once dug out, and warm spots in the wee cold hours before dawn. In the walls of the pit, we'd chisel out shelves and cubbyholes for the set proper, for the candle, for a filing cabinet and for the remote unit. Moreover, the pit must be well camouflaged and protected by mosquito netting and at night by one or two shelter halves to conceal the candlelight.

The communications command post was set up with radio, message and wire centers. Our radio station now had six assigned personnel, three on duty, three off. In each battalion set-up there were also a telegraph set, two intelligence men and a couple of buglers. We got to griping about how little work the buglers had to do in a combat area so tactical, with noise kept to a minimum. One of the buglers stayed with us for a while, and after listening to us he picked up the 284 set once on a move and carried it alone the whole distance. Since each of us had carried that set ourselves, we grew to have more respect for those following the bugling profession. I don't recall ever hearing a bugle on maneuvers, but I suppose if the command post had been overrun by the enemy, we'd have had a rout with musical accompaniment.

The miniature CPX ended Saturday, and the tactical cover was removed from all operations and we were given a day of rest. I recall waking for chow and mail call only, the rest of the time being devoted to catching up on lost sleep and trying to store some snoozes for the coming week.

We arrived at the new bivouac site at six a.m. Saturday morning. I took a well-deserved and refreshing bath in a nearby stream. Anyone who didn't soon had a wide circle of friends. Any radio operator who worked with me in a dank, closed pit covered with tent halves had to be freshly washed or he would be ushered out forthwith.

We now had indelible marks on us—innumerable mosquito welts and chigger bites. We each had mosquito nets or we'd have been eaten alive. My netting, of course, wasn't long enough. On the first night I slept with feet poked out the end and got stung. After that, I slept with combat boots on, and made use of the little bottle of Eveready Insect Repellent we had received just before leaving Fort McClellan.

Church services were held at 1230 next day, Sunday, and the great majority of the company attended. The candles and elaborate gold cross were placed neatly on a white cloth draped evenly on the hood of the Jeep. The Chaplain gave an excellent service and readings and we joined in the hymns that floated through the forest air. Afterwards, Sunday dinner consisted of chicken, mashed potatoes, peas, salad and iced cocoa. The dessert was something we hadn't expected way out here. Ice cream. We rejoiced in the luxury of those two in-between days—no camouflage, no utter darkness after the sun's descent, and nothing tactical.

Everyone's pet hate during tactical operations was that of eating. It all had to be done in the dark… breakfast at 4:45 a.m., supper around 9 p.m., sack lunches during the day. There must be no chow line, to speak of. To line up for eats meant keeping five paces apart, making for a mile-long line, and perfect conditions for chowhounds to break into line. No mess kits or papers hung on trees or lay on the ground. A strictly tactical camp required rigid discipline twenty-four hours a day.

The captain's orders fell on those indomitable spirits of everlasting ingenuity: the squad leaders. Theoretically, we knew everything. The sergeant kept us jumping by that old pat-on-the-back, I-need-you-boys method, and I wanted sometimes to swat him with a shovel and tell him to go-find-some-other-boy-you-bastard.

We had apparently picked the wrong week in history to be trapped out in Talledega, for world events were picking up considerably and we often wished for a paper or radio. An ordinary radio, that is. Word trickled through that Rome had fallen and we clung to the hope that it was true. No matter. The Battle of Talledega must now resume, with fury renewed.

At zero hour, 1800, news swept through the forest of increased action and fierce fighting on all our Talledega fronts. We felt Chateau-Thierry and Anzio Beachhead had nothing on us. And for all the snakes, bugs, and spiders, neither did Guadalcanal. So we went out to the horrible five-day conflict ahead, held together by the thought of next Saturday night back in camp.

There wasn't much difference between this and the previous week's miniature, except that now, fortunately, we didn't move as much. We went forward about once in 24 hours and usually at night. This gave us time to dig comfortable emplacements and to lavish them with seats, side tunnels and all the other comforts of home. We soon began to feel like ground squirrels. Each time we moved, I lost all sense of direction. If the sun didn't come up in the north and set in the east, then all day long it moved southward. The Alabama jungle could do that to you. If it weren't for the pines, we'd have sworn this was New Guinea.

I spent the first two days of the problem on the telegraph, TG-5, which seemed a much faster way of sending messages. It wasn't dependable, though, like radio, for when the enemy cut telegraph lines, communications went out. The wire platoon had a comfortable thought during training. They were told that in combat the enemy had a nice trick of cutting wire then hiding in ambush for the construction crew to come out and fix it.

I was sitting on the TG set the afternoon of June 6th when the operator at the other end contacted me and sent a three word message. I looked at the three words, then tapped back to ask him who he thought he was kidding. He said he was serious, so I looked back to the three words I had copied down on the yellow sheet.

INVASION HAS STARTED.

Intermittent bits of news began to pour in from the message center to confirm this little tidbit of glad tidings, and I came to the conclusion that things must be popping over on the Continent. Apparently, Ike had launched the battle of France. We heard that a wave of C-47s nine deep and 200 miles long had flown over. It seemed incredible. Big scale operations, we guessed, but it did seem like Ike could have waited for us. The landing would have been so much easier for the rest of the fellows.

Out here in the forgotten—except to the Army—wilds of northern Alabama, there weren't any newsboys shouting on the corner or in the company street. So we waited in anxiety for more good news and kept pushing back our own enemy.

The business of being fed often became deplorable. The cooks serving us were in the same part of their technical training as we. Even with the non-coms in charge, a number of meals turned out distasteful. They had to be cooked at night and served after dark, all tactically—no fires, no light, no noise. After going all day with scant food, we became ravenously hungry at supper time, and I guess we wanted too much to eat. The sack lunches, three sandwiches and an orange, compensated to some degree. We supplemented our diet with natural foods, such as nuts and berries, wherever we found them growing nearby.

"What are you doing in that poison ivy?"

The time at Talledega had memorable moments. In the evening came the call of the whippoorwill, the rhythmic cadence with which it sang filling the night and instilling a note of contentment in the night air. It rained a little one day and loaded the leaves of each tree with a tablespoonful of water, making the wearing of a raincoat necessary if you stepped out into the forest. Once during a move, we were impressed with the weirdly beautiful ways of the southern forest. I described it this way:

> *Last night there was a full moon as we moved out of our command post and down the road. It was very beautiful. The bright silvery light was shining on the leaves of the trees and causing a bright sparkle to issue into the black surroundings. We moved along in a dispersed formation—both sides of the road and ten yards apart—and the figures ahead looked as if to be eerie automatons from some adventure on the moon, bobbing mechanically up the road, making little or no noise as they plodded along. A spoon in my mess kit throbbed in perfect rhythm with my step. After a short distance we crossed a bridge and soon came out of the dense thicket onto a moor—somehow cast out of Sherlock Holmes' "Hound of the Baskervilles"—snatched from the London outskirts and placed in the heart of Alabama. It was weirdly beautiful and struck me as a scene you could only see in your mind's eye while comfortably located in your library reading some Doyle or Hammett masterpiece into the wee hours of the morning. The ground was soft and covered lightly with grass. There was a moonlit fog—creepy...and close—which seemed like a pall over the forest. This time the figures ahead were only half visible, and almost disappearing into the mist.*
>
> *In a moment we were out of it, back into reality, and trudging along a moonlit country road. It had been one spot of fog on a clear night, rising no doubt from a nearby stream... or the passing of a diaphanous ghoul.*

On Wednesday we retreated, that is to say, we withdrew strategically to a previously prepared position. Then we moved forward again at one a.m. the following morning. We had a couple encounters with the radio-telephone remote control procedure. This was a method by which commanders could talk directly to each other when all wires between them were down. It involved a complicated process that began when one officer phoned in to say he wanted to talk with another. We ask him to wait, we'll call him back when the call is ready. We get in contact by radio with the station called and they phone the officer wanted, telling him to stand by. They call us and say their party is waiting; we call our officer and tell him to go ahead. The conversation begins. The respective operators monitor and generate the signal current when their officer talks and works the anti-howl system. But after doing this twice, the officers decided they'd have to put more remote control training in future cycles. We got their talking through, but it wasn't strictly like picking up the telephone.

When time fell heavy on our hands in the evening and chow wasn't until nine we often went huckleberry picking, with the bob-whites and chuck-wills-widows calling out in the forest. The huckleberries here were very mild, so that it was nothing to down a fistful of them in one gulp. You could pick a canteen cup full of them in no time, and eat it in no time, too. They were delicious. And we couldn't help thinking of how a huckleberry pie might taste. Any kind of a home-made pie, for that matter. On one of these berry picking excursions I also picked a leaf of poison ivy and it spread across the backs of the hands. Only the greatest will power prevented scratching damage.

Friday was the last day of the problem. At 1900 we received orders to stand by the radio for the QNW signal. When we closed down, camp would be just so much closer. This order was to come by radio because the wire platoon had taken down all its lines and was policing them up. At about nine, everyone gathered around the set. Being tactical was thrown to the winds. I placed a candle on the loudspeaker; its glow lighted the tense faces that peered toward the set. And naturally, it was one of those nights when static and interference went rampant over the air waves. But finally the order came and officially brought the Battle of Talledega to a glorious conclusion.

That night we slept back at our previous Sunday bivouac and the following morning left by truck convoy for McClellan. Along the way we passed corn fields, Uncle Tom's Cabins, tumble-down shacks, and disheveled people gathered in the yards to watch us pass. Farming patches were scattered throughout the hills and here and there an old farmer lazily hoed and plowed, his straw hat beaming in the morning sun. The people waved to us; little barefoot and shirtless kids ran up to throw us a snappy salute. We saluted back. We passed some thinly clad southern belles, too, and once I almost fell out of the truck. There was a lot of stomping and wolf calling, and the girls would curl their hips and twist their bosoms so that we could enjoy the scenery. And enjoy it we did.

TEN
DISMISSED

We found it extremely good to be back in camp again, back to ice cream, candy, swimming, movies and, of course, soft beds and chigger-free huts. Then that euphoria suddenly poofed.

The 17th and last week of training was the toughest of all, especially in regard to lack of time. Class hours were extended to 1730 from 1645. End-of-cycle details, like cleaning and turning in equipment, ripped up any free moments we had. Everything turned in had to be spotless, even all our dirty clothes from Talledega, and since laundry service wouldn't solve this problem in time, we had more hours ripped up by doing the laundry ourselves. Besides that, there were night classes and review. And if you think we had finished with the basic nightmares of war in the trenches, it wasn't so.

Each trainee had to run the night infiltration course, so, being trainees, we ran it.

If we thought we'd been dirty at other times during training, that was just sissy stuff. It wasn't raining that night, but it may as well have been. It was hot. Sweat poured in buckets as we crawled the 85 yards with rifles and light packs, and every one of those 85 yards was there. Bullets and tracers whizzed overhead at 45 inches, left fiery paths, bounced off the mountainside and shot into the air like fireworks at a Fourth of July celebration. Under this barrage we kept low, very low, flat on our stomachs, inching along.

With all the sweat and dust, we soon resembled crawling mudballs. My fatigue shirt, open at the collar, began to scoop up mud so that I weighed twice as much, clothes and all, at the end than at the beginning. In the latrine afterwards, it took gallons of water to wash off all the mud. I was surprised we didn't clog the drains.

A sign in the latrine read:

If every soldier made his equipment last just ONE TENTH longer, it would equal so many billion pounds of equipment, or enough to supply three armies the size of the Fifth, now fighting in Italy.

After reading and believing that, it became difficult to watch some of the things that went on in the hut. If a small hole appeared in, say, a pair of trousers, a cutting edge was used to increase the size of the hole so that the garment would appear worn out. It would then be taken to the supply room to be turned in as salvage and exchanged for another. At other times, apparel would be purposely ripped so as to receive replacements. That disgusted me. I couldn't do much more than chide the perpetrators. I could have hauled them to the company commander, but I'd need proof and witnesses and no one was going to rat on his buddies. Especially those guilty of the same crimes, lest they be ratted on.

This taught me Enduring Lesson Number One: *A considerable percentage of mankind doesn't give a shit.* So I let it go at that.

On the evening of June 15th, Infantry Day, the customary parade was held. During the last week, I was one of seven selected for interview for Officers Candidate School, but this was one time I didn't try particularly hard: I could think of more pleasant jobs, and besides, I wanted to make friends with people. I felt that as an officer, I'd be court-martialed for being too lenient with the men, for I never condoned many of the tyrannical methods some officers applied regularly. The best we met in McClellan was fortunately our radio officer, a Lieutenant Magne, a considerate civilian at heart. I

didn't think I had officer qualifications yet, anyway, or that I was going to make Army life a career. Three of the seven were selected for further screening on the basis that they weren't young sprouts and had a college education. As for me, I was just ready for advanced radio.

At last, the orders came for thirty of us to go to Fort Benning, Georgia, for advanced training, and needless to say, we were decidedly pleased. We heard of one fellow who took training there, and his chief comment was: "This *can't* be the Army!" That was good enough for us.

On Saturday, the 17th of June, our cycle ended mid a roaring cheer and storm of flying helmets as the sergeant shouted: "**Dismissed!**" It was the moment for which we'd waited these last four months. For the next six days there was little more than turning things in, getting packed, swimming and waiting. During the whole 17 weeks, the only item of equipment I lost and had to pay for was the wool-knit cap, price 45 cents. Someone lost a microphone at Talledega and we each chipped in 16 cents to pay for it.

We received our MOS training certificates that week:

FIFTH TRAINING REGIMENT
Infantry Replacement Training Center
Fort McClellan, Alabama

CERTIFICATE

This is to certify that

Private Myron D. Sutton

has satisfactorily completed, during the period from 20 Feb 1944 to 17 June 1944, the schedule of specialized training prescribed for a

RADIO OPERATOR ssn 776
(16 words per minute)

Joseph L. Smith
1st Lt., Inf.
Senior Instructor, Radio School
John L. Beaver
Major, Infantry
Commanding 15th Battalion

We said our fond farewells to the few who weren't going with us and silently wished them some extra luck. After their furloughs, they would report to the Port of Replacement at Ft. Meade, Maryland. That was the last step overseas. Then on Thursday the 23rd of June, 1944, we departed from Ft. McClellan, Benning bound. There were few sad tears on departure, but we did look back on the place as one of pleasant times, and a few bad, of valuable training… and the first lap of our Army career.

"Giving the place a little send-off, eh Griggs?"

ELEVEN
BENNING

We went by train, leaving McClellan in the morning, and made it to Childersburg, Alabama, where we had a five hour layover. This suited us fine because there we encountered one of the finest USO's we'd seen. Moreover, it was air cooled—a haven of delight on one of those cruel, hot days. We lounged, read, wrote, ate, recorded a phonographic disk to send home, and managed to fill the time available.

After that, on across more of Alabama, through much the same country of farms, forests and magnolia trees that we'd encountered at McClellan, Morrisville and Talledega.

Then, at last, at 12:30 a.m., Eastern War Time, we arrived in Columbus, Georgia—just across the Chattahoochee River and Alabama line.

We loaded aboard what we termed a cattle car, and rode the ten miles to Fort Benning and pulled up in front of our barracks. In the darkness, we couldn't see much of the fort, but what we saw of our barracks was impressive.

By the headlights, we noted that here they weren't huts, or even wooden frame buildings. They were brick, stone, cement! They towered up at least three stories and were quarto-circular in arrangement. On each floor, a terrace extended the entire two-block length of the building, the front adorned with impressive archways. We pinched ourselves a couple of times to wake up, but found the dream was indeed real.

Inside, we found ourselves in a hallway from which steps led upward into the darkness of the upper levels. Bulletin boards lined the hall. The guide took some of us up, led us down a long row of bunks until we found several at the end that weren't occupied.

"Here you are," he said. "This isn't really permanent, but it's a place to sleep until morning and I know you're dead tired from your trip. We'll straighten things out and issue bedding in the morning. Good night."

Such politesse! We hadn't been spoken to so politely for months, but I think we were too groggy to appreciate it. Anyway, the bunks here were no longer than those anywhere else, so I resolved to sleep scrunched up, as before. We threw our stuff on the floor and headed for the latrine.

We thought, after these last few minutes, we'd be prepared for anything, but we weren't for the shock that came. The floor was tiled in delicate orange, as were the walls. The toilets were divided and lined with huge black and white marble slabs polished to radiance. There was a tile clothes room or anteroom for the tile-and-marble shower beyond. We couldn't stand any more of this so we performed our necessaries and made our dazed way back to the bunks—some twenty steps away—snatching a drink from a fountain on the way.

"Everybody up!"

Next morning we awoke and looked around. We were in a huge room lined on both sides by double-tiered bunks, and a double row of lockers down the center. Men stirred everywhere, brushing up for the day. We looked out the window and there, spread out to the horizon, grew green trees as far as we could see, dotted by red tile roofs. We got up, washed and dressed—still thinking this kind of thing was too good to be true—and went down to breakfast.

The dining hall ("mess hall" became an obsolete term; dining salon perhaps would have been more appropriate) was also in modern brick and tile. The kitchen was spic and span, and breakfast excellent. After filling up, we lounged outside to watch some men drilling on the parade ground. There was a large, open field in front of our building and another identical building across on the other side. We now saw in the day light that our barracks rose four stories. I thought the landscaping of lawn, trees and shrubs very attractive, unless, of course, I had to mow the lawn.

We watched the drilling. Uniform: fatigues, with the shirt tails out—for ventilation, we assumed. No leggings. And instead of rifles, to which we had become greatly accustomed, these people were going about with books and manuals in their hands! Had we taken a wrong turn and arrived at a college? We couldn't believe that something wasn't wrong about this. We didn't fit. But we could learn! And we instantly began to like it. We figured we'd like all three months of it, too. That morning, Sunday, was spent organizing the company into formations and listening to a talk from our company commander, enlightening us on the things we wanted to know.

We received a copy of our schedule. Training would begin promptly the following morning at the ungodly hour of 0545. It would consist of preliminary tests, then code practice, aerial photo and map reading, electrical circuits, vacuum tubes, radio waves in transmission, message center procedure, cryptography, and wire. To us, the schedule, at least for the first week, looked good, the variety adequate.

We got off duty at 1245 and had the remainder of the day to ourselves. One of the boys said: "From the looks of this, fellas, we're sure gonna get soft." Came the reply: "Who in the devil cares?!"

We moseyed around and gawked and asked questions. Instruction would be divided into two time periods: 0545 to 1245 and 1315 to 2015. We would alternate two weeks with the morning shift, then two weeks in the afternoons. The company, known henceforth as Enlisted Communications Class Number 50, or ECC 50 for short, was divided into four platoons: 1st from Camp Walters, Texas, 2nd and 3rd from Camp Blanding, Florida, and 4th from Fort McClellan, Alabama. We had entered a very much bigger world!

Most of the company slept in a large room accommodating about 100 men. There were laws laid down for us, too: no dust or trash on the floor at any time, everything maintained neat and clean, no scuffling or loud carousing, lights out and bed check at ten o'clock each night.

Our barracks building, or dormitory to be polite, was an integral unit. It had its own barber shop, mail room, supply room, writing room, day room, and study hall. The writing room had eight modern plastic black-topped tables and upholstered spring type chairs, with fluorescent desk lamps. Such elegance popped our eyes wide open.

On the roof, we were told, there was a sun terrace. Several of us put our gear away in foot lockers and wall lockers and then headed off toward the swimming pool. On the way, we stopped in at the Main PX to down some ice cream and do a little shopping, then went on—passing a magnificent theatre building and Service Club. The swimming pool was as magnificent as everything else around here, with diving boards, walks, a slide, and large enough to require five life guards. This was the Army?

We found that we wouldn't lack for facilities for exercise. There was a gym nearby and places for tennis, badminton, bowling, baseball, horseshoes, and almost any other recreation we could bring to mind. If this amounted to a playboy's country club rather than a military establishment, no one protested. Until next day.

Thus, then, did we start life with the 24th Company, 1st Student Training Regiment, ECC 50, Ft. Benning, Georgia. We'd liked McClellan pretty well, but this—this was too much to take all at once. We thought there wouldn't be anything to gripe about, but our grumbling began the following morning when we were roused out of bed at 0330 to begin our first day of classes. At length, we got used to the hours and regulated our activities to correspond therewith. Since things were so convenient, we showered every morning and again in the afternoon after baking in the heat up on the sun terrace, where many of us did our homework. Getting up two hours early gave us time to wash, eat, clean up and have an inspection—perhaps to cram a little, too—before class time. When classes got started we normally had quite a load to carry: six manuals, five maps, workbook, pencils, protractor, coordinate card, and notebook. Just school kids, that's all... just school kids.

I had a friend at home, named Scott Colby, who had joined the Navy several months before I got into the armed services. Naturally, he boasted at the time that by going in first he would get across first and so do more fighting for his country. This turned out to be a misconception of anticipation, for at the time I came to Benning he still sweated it out in the V-12 program and by that time was getting fed up with going to school. Even some of my basic training fortunes intrigued him. He got into electrician training, which wasn't his cup of tea at all, and began to profess a bitter envy at my seeing the country and having a good time while he sat at home. Our little intra-service rivalry became acute later on, as indeed you shall see. His mother sent me a box of candy at Benning and included a note saying: "Quit your crabbing because it isn't full—I had to send some to Scott, too!"

Benning was the Infantry's chief parachute training center. From our windows we could see the parachute towers where trainees were lifted in open chutes then let go, a process we assumed would give them the final details of the drop and the landing before going up for the first solo flight. Most of every day was spent in chutes going up and floating down. Crowds of leather-jacketed or fatigue-clad paratroopers waited in line below. Those who weren't actually in the jump at the time spent their free moments at calisthenics.

They said the training was rough. To find out just how rough, several of us went over to the paratroop barracks and found our old friend Scott Foss of Hut 9, who gave us the low down. In his tough outfit, discipline meant everything. The paratrooper, he claimed, had the brightest shined shoes in the Service. By looking at some of them in the barracks we didn't dis-

agree. Foss's company quarters were simply frame barracks and he didn't want to hear how we plutocrats got along in the brick-tile dorms with the sun terraces. Nevertheless, he found his training interesting, albeit rugged. It only lasted four weeks. They jumped off the towers in the third week and out of planes during the last week. Preliminary training involved jumping parachuteless off towers ten and more feet high. Foss was going to try for paratroop radio after this basic stuff and it seems to me he made the grade.

"Now what?"

Just why the odd pastime of hypnotism should develop in this particular paratroop company we never exactly found out. Foss said one of the boys was particularly good at it, and even Foss had learned the technique himself. He came up to our barracks one night and we got in a private room just off the barracks or squad room and found out he knew what he was talking about. He put Bob Hill and Aubie Darr completely under. I tried it out, and my smooth, glib voice put Darr out. Or else he just fell asleep. They say hypnotism is a study of the mind. When they tried it on me they failed. After all, they have to have something to work with.

After two weeks, we changed to the afternoon shift, and we liked it somewhat better even though it meant readjusting our sleeping hours. At that, we got more sleep. However, inasmuch as it meant two hours of classes after supper, that dispelled the idea of getting into Columbus in the evenings. Kept some of the boys out of mischief, though.

Free 16mm movies, shorts and feature films were held outdoors from time to time in front of our dining hall. This turned out to be a welcome way to spend an evening because shortly after arriving at Benning everyone went broke. The nicotine addicts became weak and made continual fools of themselves bumming. It got to the point where any lucky soul who stumbled somehow onto a pack of tobacco sticks had to smoke them furtively in private. I gave up smoking until payday, though I did have some tobacco left for my pipe. I had to laugh, though, at some of the nicotine fits that went on—where the strongest weakened and the weakest just about dropped dead.

The food continued top choice. One typical Sunday dinner consisted of fried chicken, corn-on-the-cob, string beans, mashed potatoes and tempting brown gravy, salad, hot rolls smothered in smooth, rich butter, several cups of ice cold lemonade, and cherry or black walnut ice cream. We never had it so good. Frequent bus trips made it easy to get into Columbus and I made several trips. On one of them, we saw a woman carrying a basket of flowers on her head, the cranium being one of the commonest modes of transportation in that region. It wasn't a small basket, either, by any standards and we looked for it, flowers and all, to go plummeting to the sidewalk. Obviously, she had perfect balance, curbs, corners and all. Trademarks like these reminded us now and then of the Old South.

During our morning classes, we would frequently spend the time around six a.m. in Building A, which just happened to be across the street from WAC barracks. About half an hour after classes had begun, someone would give a suppressed whistle and heads in our classroom automatically shifted to the right. I mean, if you have the Womens Army Corps as a neighbor, you should take an interest in your neighbors. On the drawn shades, we could perceive shadows of the young ladies performing their morning rituals, a subject far more interesting than the radio concepts being monotoned by our instructor. During the pursuant ten-minute break, the WACs would be out on the parade ground doing their morning exercises, and we would stand around staring at them, each of us with a stupid expression on his face, drooling, and with tongue hanging out.

ENDURING LESSON #2

"Human beings have two basic passions: self defense and reproduction, not necessarily in that order."

You may think this improper behavior and indicative of bad upbringing. But you must face what I had learned as Enduring Lesson Number Two: *Human beings, like a million other species of animal life on earth, have two basic passions: self defense and reproduction, not necessarily in that order.* The pursuit of the female by the male (or vice versa) has been going on since before the days of the trilobites, 500 million years ago, an evolutionary trait that persists to modern times. Persists and intensifies. As the evidence suggests, human beings have become oversexed. There are some people who would like to see evolution reversed on this point, but a majority prefers to yield to the pressures of procreation. Suppressing these pressures is abnormal. We were just being red-blooded American boys.

The Army got enough of this, however, and the following order came down:

Ogling of WACs is prohibited.

Shortly afterwards, I noticed a sign in the supply and repair shop that dealt with the WACs on the one hand, and with communications personnel on the other. It ended this way:

"...Furthermore, the sending or receiving of flashlight signals to the windows of said barracks is strictly prohibited."

Whoever said we wouldn't have inspections at Benning was full of whatever people like that are full of. We had 'em, and in quantity. The powers that be warned us that an inspection could be held in our sleeping quarters at any time we were away. Eventually it dawned on us that no such inspections were being held and we relaxed. This we should not have done. One morning a surprise inspection took place and it turned out the way all such inspections do. The captain not only found books, magazines, newspapers, clothes and towels on bunks, but discovered complete disarray in foot and wall lockers.

He called us out and gave the blade to us in no uncertain terms. The place was to be cleaned so well (by us), and placed in such perfect order (by us), that it would pass a General's Inspection. It would be inspected with a fine-toothed comb while we were at classes, and if it wasn't perfect we would begin cleaning and scrubbing the barracks every morning after that and there would be no more passes. If anyone got caught not doing his share of clean-up, he would be obliged to some extra time in the kitchen—KP in other words, the Army's capital punishment.

This emphasis on extra work was fine, a necessary evil if we couldn't manage to keep Uncle Sam's fighting forces clean. But we had to share our buildings with another army with which little attempt at combat was made... cockroaches. They were big and strong cockroaches, too, bred in the wholesome climate of west Georgia. I heard a couple of them one night discussing me. "Give me a hand in carrying this guy outside, will you?" "Naw, you take him alone. I'll bring the bunk."

They accepted a few more at Benning for OCS, Officers Candidate School. If any of us had intentions of becoming radio officers, they were dispelled one evening in an informal chat with one of our training officers. Someone asked him what were the chances of an officer in front line combat as compared to a radio private. His answer was as quick as it was definite.

"You men haven't been in the Army long enough to be soldiers or you wouldn't have asked such a question. The chances are 600 to one in favor of the private. When he lets out a peep from his radio and shells start flying, his urge is to get the hell out. The officer must stand up there with his carbine, shells flying all around, keeping the privates from jumping out of their foxholes and getting shot."

Suddenly, we all wanted to stay privates.

I became a squad leader again here at Benning, though through no choice of my own. It wasn't bad, however, and didn't require extra work. We met a new bunch of fellows there, too. Henley, from Ithaca, New York, couldn't believe in Arizona being anything but desert until I showed him a couple copies of *Arizona Highways.* There was Joe Vilanaro from Boston, John Reinhart from Ohio, Vic Roberts from New Jersey, and Beaumont Gunn from Memphis. Beau Gunn, as we called him, was still steadfastly determined that the South never lost the war to the damn Yankees. The Yankees present defended their positions, and verbal free-for-alls followed in the squad room. Gunn's grammar was Deep South: heah, theah, and everywheah. Pronounced hee-uh, they-uh, and everwhey-uh. He claimed his home was on the Powdah Rivah.

We learned with astonishment that Ft. McClellan's communications school was dropping by the wayside. Of eight radio companies, only one remained. The rest had become rifle and heavy weapons companies. Apparently, we had arrived there at the right time. I had wanted to get into radio repair school after finishing this tour at Benning, but never made the grade. It was just as well, though. The code began to get us in time. One day some of us actually laughed at this one: Said the baby ear of corn to its mother, "Mama, how did I get here?" Reply: "The stalk brought you."

We got word about some of the boys in Hut 9 who hadn't come to Benning with us. Winnie Morgan reported from somewhere in Italy, saying Walt Garrison was with him and for us not to pay any attention to all that bull about the horror of going overseas—he was having a grand old time. Said he'd seen the Volturno, the Isle of Capri and Naples. Kid Harrick, Andy Hochstetter, Gil Bower and a number of the other C-15 boys were in that area. Jimmy Ruff had landed in Mississippi. They had been dispersed to all corners...

But life at Benning went on and we lolled about in lumbering ease, like hippos in an African river, a placid life. But we also worked hard and learned some heavy stuff. Radio became an advanced subject, and we loved it.

TWELVE
CLASSES AND CODE

Advanced technical training at Fort Benning turned out to be one of the most interesting periods in our respective Army careers, at least for those who like learning, and lots of it. Life was easy, and carefree, though we were immersed in oceans of work and study. We enjoyed the advanced work, the deep challenges, the applications we made of our previous training. Everything was on a big scale and we were given the works. We wore fatigues always, had breaks every hour and lounged around on our time off. Many of us did our homework up on the sun terrace and began to look more like Indians than whites.

At first we had code aptitude tests designed to place us on the proper tables that suited our individual code speeds. The boys from McClellan shot skyward and pulled away from the men trained in other camps. Fellows from those places had been able to pass the remarkable speeds of 22 and 24 words per minute and now couldn't pass 12 wpm. The reason was simple. Their training had involved nearly all clear code groups without numbers, a snap in any radioman's language. Our practice had always included mixed letters and numbers. Consequently, we pulled into the general lead. Speed progress was posted on a bulletin board at one end of the code room by means of name tags in the proper columns. Upon passage of a certain speed, our names were advanced. I got through 12 and 14 wpm on the aptitude tests June 29, 16 wpm the next day, 18 wpm July 8, and 20 wpm on July 20. All in longhand, remember. This, I thought, would certainly wind up my code career at Benning. When I picked up the earphones and listened to the 25 wpm speed, I knew I'd reached my peak. Nevertheless, I began to struggle with it, and from then on, hour after hour, day after day, in the code room, I worked mightily to pass it, along with three or four other speed demons. Some day, I said to myself, I might pass this. But then, that's the stuff dreams are made of.

When it wasn't hot it was raining. Yet here we didn't notice so much because we were never out in it, that is to say, not on maneuvers. Occasionally, however, a downpour would begin on our march to or from the class area and we would scatter to the sanctuary offered by trees, doorways, and buildings. The rains in Georgia could be just as torrential as those in Alabama and could last just as long. The streets ran like rivers and we often got drenched before we could even start for cover. The rain cooled things off and we loved it. Sometimes fog cast a weird effect over the fort as viewed from our second story windows. Fog also brought oddly beautiful mornings when we arose at dawn for the morning shift.

The main post theatre was a beautiful, air-cooled place, though at times I figured they could have used a better operator. I would gladly have offered my services as projectionist. Especially on the Lana Turner or Betty Grable offerings.

Course schedule showed 205 hours of code practice, 128 hours of field nets, 67 hours of radio sets (function, installation and maintenance), 42 hours of radio procedure, 19 hours of cryptography, 16 hours of wire communication, 14 hours of map and aerial photograph interpretation, and 8 hours of electricity. Here we delved into the intricate designs of radio circuits and though we got lost from time to time, found it not a little interesting.

The first week of training, we got a taste of how magnificent the Georgia countryside can be in full summer. It wasn't that much different from Alabama, but it was warm. Trucks took us five miles beyond the fort to make position sketches, a part of map training which taught us the principles of recording field observations topographically. We walked down one side of a small valley and back the other side making rough sights, directions and distances and drawing in roads, wire lines, and other cultural objects. The same day we had a GT (Graded Test) in map reading and I made AA on it. This was the last time I was able to boast of being highest in the company. When the going got rougher, later on, I fell back down to where I belonged.

A couple of us walked over to the Chattahoochee River one day, just a short distance from the fort. It was much larger than I'd expected, some seventy feet across, shallow and muddy, with thickly wooded banks. We found a large, brown lazy turtle on the bank and flipped it off into the water. As we started back, rain began to pour, but we caught a post bus in the nick of time. I heard once that someone put a small newspaper item in a Columbus paper about the battleship Missouri having sailed up the Chattahoochee and run aground a short ways above Columbus. Since the Chattahoochee wouldn't support a kid's sailboat, this item was an experiment to find out how gullible the public could be. It worked. They reported that 600 people drove out to see the battleship Missouri stuck on a sandbar in the Chattahoochee River.

One of the most impressive parts of our training was a movie on radio transmission security. The plot concerned two task forces attacking the Germans by sea—one achieved great success while the other almost failed. Why? It all began in a town not far across the water from enemy occupied territory. A radio operator had been ordered to stand by, not to transmit and not even to receive, since radio tubes themselves sometimes emit radiations. But this particular operator decided he couldn't do any harm by taking off a couple sections of antenna and sending low power to a friend some distance away in order to ask about shacking up with a certain gal in a city they'd just left. Of course, an enemy intercept station picked him up and discovered where this truck convoy had been and figured out just about where it was now. So… bombers flew over and practically wiped out all the trucks, with a consequent loss of numerous lives.

The other elements of the task force kept on, however, and some of its bombers flew over German territory in advance preparations several days later. The static and interference over the air was terrible and the radio operator in one of the planes arrived at the point where he could no longer take it. He moved the phones away from his ears and onto his temples, sat back and began to take life easy. An order came shortly for the flight to change direction because it was heading directly into a squadron of enemy fighter planes. All the planes received this order and veered away—all, that is, except one—whose radio operator just couldn't stand that horrible loud static. So his bomber, in a cloud bank, failed to observe the others turn and thus flew right into the enemy fighters and was downed under heavy opposition. Lost: one plane, the crew, and millions of dollars. All because one radio operator couldn't obey orders.

The task force went on, depleted now, but nonetheless ready to carry forth. The convoy sailed. A radio operator on one of the ships had to make a transmission by code to another ship. He had grown lax and let his instruments drift off frequency. Not knowing this, he began to send. A pack of Jerry subs waited nearby, trying to get a bearing on the convoy. Suddenly a sub operator picked up the off-frequency code and listened. The ship's operator got an answer to his message, then. A German answer. All at once, he discovered he was off the beam and that the answer couldn't have come from the other ship. Before he'd straightened out his dials, the torpedo struck. The radioman on the enemy sub had heard the call just long enough to set out his direction finder and head the wolf pack in the direction of the convoy. Loss: three ships, their valuable cargo, and scores of men. Reason: one radio operator.

At length, the convoy attempted to land on the enemy coast. But now, of course, the Germans knew they were coming and were well prepared to defend their coastline. With blood and sweat, however, the first men ashore established a beach head and the fighting continued at a fierce rate. In the heat of one of these battles, a radioman was required to send a message by voice and found considerable difficulty in making the party at the other end understand the word "Devil," which over the air may sound like rebel, devoe, revel, and a host of similar words. The sending operator, mad at the other operator because he couldn't understand such a simple word, launched some cusswords and gave away his headquarters position. This the enemy picked up and began to lay down a heavy artillery barrage on those particular buildings and a nearby bridge. All because one radio operator failed to make use of the phonetic alphabet.

An enemy machine gun nest mowing down hordes of onrushing soldiers prompted an order to a radioman to call for artillery support. In his excitement, the operator attacked the key and sent at a furious speed no other station operator could possibly copy. They asked for a repeat and while the operators fumbled, more men hit the dirt with machine gun slugs in them.

The other attacking task force, meanwhile, made a successful landing farther down the beach and drove inland. Because, of course, they had such excellent radio operators. Eventually the two forces met and the battle was won. The End.

I didn't notice at the time, or else I've forgotten, but this saga could have been directed by Frank Capra or John Ford, on the heels of their directing such peacetime epics as *Lost Horizon* or *Stagecoach*, respectively. Whatever the case, professionally well-done Army pictures impressed upon us the reasons for becoming radio operators who never broke the rules. That way, we could have every reason for expecting a long and happy life. So we bore down in earnest and began to listen in class, for a change.

We learned more about the radio sets SCR-536, -511, and -300. We had some instruction in the Signal Corps pigeon loft, observing the art of sending and receiving messages by this air-mail method. We were shown several types of birds, one of which was better trained than its playmates and put on an impressive acrobatic show for us. We continued with message center procedure and gained more hours of practice with the M-209 converter, still classified Confidential. We got sending practice in the code room so as to develop "the perfect fist."

Sometimes roll call of the 210 names in the company permitted us to catch up on sleep or write a letter home. Enduring Lesson Number Three: *if you look carefully, you'll always find a way to loaf.*

R-T Procedure, or radio transmission nets, were practiced on the code tables from time to time. Here we got most of our practice in the radio shorthand of Q signals. QMC: I am moving and will endeavor to keep in touch with you while on the move. QNW: I am now closing my station and leaving the net. There were many others, QMA, QKG, QYT, QQZ, QJS, and so on. We added one when the fellow on the other end began fouling up. It was QQQ: Wipe the bird shit off your antenna!

The classroom instruction, believe it or not, had a great many memorable moments. The instructors, some of the best in the Army, regaled us with tales, used unforgettable instruction techniques, and helped us to laugh in our very serious work. We'll never forget Sergeant Crabtree, a tech sergeant who handled much of our classroom work. His instruction was perfect in all the essence of the word, and none of us has probably ever had such effective teaching. The fact that he added humor to all situations took the dryness out of the subjects he taught, which, under someone else, could have been as dry as the California desert.

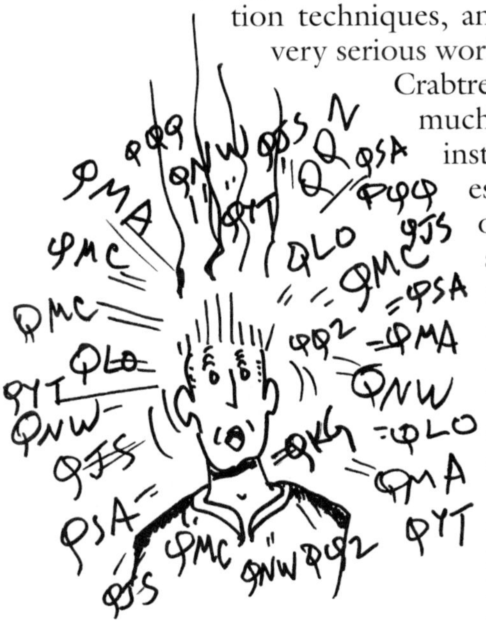

Q Signals.

He vowed to give us a logical reason for everything he taught and would never have to say: "The book says so." Once in a while, someone would ask a question that trapped him, whereupon he'd step over to the side of the stage and hide behind the curtain. After a few seconds he peeked out at us and gave the only answer he could think of: "The book says so."

Few ever fell asleep in his classes, but when some did he had an ingenious device for waking them up. He would stop in the middle of a sentence and shade his eyes, looking out over the audience. Then he would pick up a piece of chalk and wing it clear out over the crowd toward the offending dozer. And Crabtree had remarkable aim.

One day, he said: "This next question is one which I ask and always lay odds of ten to one it's answered wrong. I usually win. Question: What is the reply when you get an IMI K on a message?"

He glanced at the roster before him. "Ah, who's going to be the lucky one to… Private Sutton."

I stood up and answered. "You repeat the message."

"Oh, HO!" Crabtree said. "Who wins the odds? You, sir, Private Sutton. Yes, indeed. That was fine… fine. Where'd you learn that?"

"Fort McClellan."

"At Fort McClellan?" he responded. "Yes, I used to live in Anniston… Hah! Indeed, I could tell you a lot about that place… Yes, sir. Dining, dancing, and knifing nightly."

Just when we started on the rougher parts of the Radio sets 211 and 284, Crabtree became ill. They replaced him with a Warrant Officer Holbrook, decidedly a contrast to the sergeant, and whose grammar was atrocious: "Do he send this message? Yes he do!"

On August 3 we saw *Mister Winkle Goes to War*, the picture Eddie Robinson and I had filmed back at Fort MacArthur. It seemed they'd put in most of the scenes of him and left me out. This I could never understand… such good talent gone to waste. While at Benning, we learned that General McNair had been killed overseas; he had visited Fort McClellan our last day there.

Code practice continued and we started speed work for record, this time in codez—code without numbers. I worked my way back up to 25 wpm. It still seemed just as fast. Having to take three perfect consecutive minutes out of five was indeed difficult—and I think many would have passed it if they could have written that fast. It's a business in which one cannot be ham-handed. Sit down some time and have someone read off 25 five-letter code groups per minute for three minutes, and print them down in perfect capital letters. If you don't get writer's cramp, you may produce a few tortured, illegible letters.

We had practice in jamming and interference and how to overcome it. And always came back to code. One day we had seven consecutive hours of it. We then began to realize why radio operators were the Army's Number One scarcity. They got Section 8s (i.e. went mad) from taking code.

This made it refreshing when our training occasionally veered away from radio. We received preliminary instruction in the field switchboards BD-71 (six lines) and BD-72 (twelve lines), their installation and operation. Then we went out to Glen Field for abbreviated wire nets—short line operations simulating battle tactics and battle moves. Visual signalling, we learned, had a definite place in communications. During the first hours of the morning shift, while it was yet dark, we practiced sending code by lamps and at the end had to do it officially at five words per minute for the coveted MOS 740 medium speed operator's license. Other visual signalling involved the use of panels, their designs and colors, and their display under tactical conditions. This, in combat, was a good way to get shot. We much preferred the lamp system if we must have any visual work at all, because our lamps were equipped with key and cord that would permit sending the message from a foxhole and prevent an embarrassing situation should the enemy decide to take a potshot at the light.

The program became so varied that I fully expected instruction in the principles of smoke signals. Eventually, it all came down to the hidden radio set as the most effective way to communicate during battle. Even so, the instructors constantly reminded us that enemy direction finders might be able to pinpoint our location and send in a shell to blow the radio sky high. A polite way of saying that we, too, would get blown sky high.

THIRTEEN
SPEED DEMONS

Com A was Radio Lab and classroom. Com B was Wire Lab and classroom. Coms C and D were both classrooms. Coms E and F were code rooms. None of the buildings was air-cooled, though fans of considerable dimension, set in the corners, increased air circulation. Walk in front of one, and you'd be set off balance. Seating capacity of the Coms and Coderooms was slightly over 200, thus necessitating a loudspeaker system for class instructors.

The two code buildings were directly connected by an open, wired-in room that contained a vast array of code recording and sending machines. Here, the instructors kept the respective tables supplied with proper code speeds. Usually, one building would be devoted to the simpler lessons. Now and then, a group of officers would come in for practice, but we learned that most of them needed only to pass 8 words per minute.

After each class period, we would lounge out on the sidewalks by the buildings and smoke, talk, and drink Cokes or Pepsis dispensed by several machines. A cold drink resulted in fresher minds after code practice. Yet how could we fight the lullaby effect of higher speeds? In due response, my head nodded. Here, as at McClellan, the cadre used effectively a familiar wake-up device: stacks of code pads dropped on the desk beside a dozer's ear. This "cannon burst" seldom failed to wake the offender.

On August 22nd Class 48 graduated. In two weeks class 49 would be out, and then, two weeks after that, Class 50... us.

It turned out that Class 50 had the highest code speed record ever achieved in Benning. Oh well, you take a bunch of geniuses like us, and that sort of thing is bound to happen. Call me Speedy.

We had a couple hours on the movement of command posts—how to shut down a station, disassemble equipment, pack up, and move to the next site. The instructor was a lieutenant who had been in the Sicily and Italy campaign. He described the Anzio beach head and how the whole line was dug in. He said one of the reasons for the comparatively slow progress at the beginning of the Normandy campaign was that most radios were silenced or not used at all because the authorities feared the enemy could pinpoint them with direction finders, and smash them with artillery fire. At length, this was discovered to be nearly if not entirely false and the use of radios not only came into being again but was considerably stepped up. They determined that the enemy didn't have equipment accurate enough to pinpoint our location. General Bradley issued the order that exploded all previous fears on this and asked for increased radio communications. After that, the campaign advanced remarkably.

They didn't spare the cost in training us at Benning. When classroom instruction finished, we went out in the field and observed battle set-ups and their operation. One was out at Gordon Field, where we sat in stands and had the field before us described in detail, point by point. At the break I went down and talked with one of the artillery men. He explained and illustrated how he brought the big guns into play on an unseen target by use of a map and various scales—and how he utilized radio.

A liaison plane flew overhead and took part in the demonstrations. It dropped a parachute flare once that started a small forest fire, but this was quickly put out by the assistants. We got a brief glimpse of how dogs carried messages. Shortly after this, we began spending a lot of time out in the field in actual radio nets. For these exercises, we used the SCR-284, which had played such a vital role in the Battle for Talledega. Once or twice we got in communication with and sent messages to a circling light plane and the following day to a B-25 bomber.

Our first day on field nets started out delightful. We set up under the protecting roof of a tennis court bleacher. Wise decision… for it began to rain. It was all comfort and convenience, too, for we sat back in the seats and operated in royal style. Not five steps away stood an ice water drinking fountain. The best thing perhaps was the view this height afforded of the officers' swimming pool at the Officers' Club. While we were there, a delightful prancing sea nymph tripped out for a plunge—bosom held high in an offensive attack and hips squirming in preparation for the dive. No restrictions on ogling that.

In spite of such distractions, the net operated efficiently. It did us good to get out in the Georgia countryside after so many hours in class and code rooms. They made a movie of us once, an entire sequence involving getting and checking our equipment, loading up and leaving the post, then arriving in the forest, setting up our radio stations, and commencing operation. The photographers said that the film was to be part of a pictorial guide to The Infantry School.

Usually, we had to crank a generator to power up our radio set and transmit. Now and then, we operated frame sets driven by a power converter and storage battery, making operation much easier. We stayed out in the field through supper when we were on the afternoon shift, and this involved sack lunches, more like a picnic than anything else. Some days it rained and some days we got soaked, but mostly we huddled under the canvas top of our truck and poked our arms out occasionally to send a message.

All this may sound like a country club at military expense, and I will concede that the surroundings had merit. But every comfortable hour was filled with intense cerebration at the radio set, review of manuals and Q signals, listening for messages, and sitting like a panther ready to spring on the next message and transmit it—correctly and at high speed.

Besides, however beautiful the fort may have been in late summer, insects, gnats, no-see-ums, flies and chiggers flourished along with the flowers and foliage. These pests buzzed into our ears during code work when we tried desperately to pass a test. Or when we stood rigidly at attention on the parade ground with the colonel nearby. If you have ever been prevented from scratching your nose at a crucial moment, you will know the anguish and torment.

Nevertheless, we communed with nature considerably in our field net areas. Pines graced with vines lent the woods a fragrant scent. From high spots on our routes we looked out over forests to the far horizon, broken in places by a canyon or deep ravine. The winding Chattahoochee made its muddy way to the south, like a ribbon of brown cutting through the green. The Upatoie, or yellow-green meadows, and the red-tiled roofs and cream yellow stucco buildings punctuated the cool green of the land. The only smoke came from a passing train, or from some industrial plant in Columbus, Lummus or over on the Alabama side.

When we went out on field nets, we carried envelopes that contained signs reading, in large red letters: DANGER, DUD! At first, we thought that these were to hang around our necks. It turned out that prowling about in the Georgia countryside, as we did, might bring us upon an unexploded artillery shell. We were requested merely to place the dud tag nearby and not toy around with these little packages of sudden death, lest we come back without a head and not much of a body. Other bundles of death existed out there, too—rattlesnakes and copperheads—and although we never encountered these reptiles, they were always present in our vivid imaginations, rearing up and lunging forward.

After a great deal of practice, our field nets pepped up with the spirit of competition. We were allowed to choose our buddies and set up our own net, then go out and send as many messages as we possibly could and as fast as we could. In this way, we vied for the coveted trophy: membership in the QLO-Left Foot Club. This code slang meant: "You're not doing worth a damn sending with your fist, so you might as well change to your left foot." A group of us speed demons from McClellan teamed up and burnt the airways sending 41 messages in four hours, and receiving messages sent to us by other stations in the net. The code key and the radio generator almost began to smoke. We didn't win, but had a barrel of fun. We might have been in line for something like QLO-Right Elbow. Few people are as nuts as radio operators.

In the last days, code practice tore me to pieces, since for days, it seemed, I came within a fraction of passing that great goal, 25 words per minute. Each time, I handed in a paper I thought to be perfect—a full three minutes of perfect copy. By September first, two others had passed it and it seemed that half the class was struggling to do so. But by the evening of September 1st, I'd had my day, and it was a red letter one. I became the third to pass it. This was glory hallelujah, and put finis on speed. If anyone takes code at a faster speed, it is done with typewriter. So, then, I sat at the exclusive table and copied clear text, that is, the kind of copy you read, say, in a newspaper. This proved to be exceedingly difficult for me, since when copying it down I became too engrossed in what was being said to keep up with it.

That same day, things really got hot for Class 50. We were restricted for a day, had to double time to class and back, had lights out an hour early, and got a grandiose chewing. It all began at chow. My platoon was the first to eat that evening, and I'd led my squad inside when, for some unknown reason, the other three platoons outside were suddenly dismissed. Ninety men stampeded to the door and tried to go through it at the same time. Twenty newspapers on a stand just outside the door were destroyed and the solid plate glass door itself broken off its hinges.

Next morning, we fell out and drilled and double-timed an hour. The captain probably thought it punishment, as so many captains will often deceive themselves, but it did us good to have a little exercise. We passed Saturday's inspection, and the restriction was lifted. The door was fixed and everyone forgot about it. I doubt if the war effort had been hampered in any way.

Naturally, with the closing weeks of the course creeping up on us, rumors began to fly... thick, hot and heavy. Where would we go? What would we do? Radio Repair School... Port of Replacement... Infantry Divisions... these and other destinations were bantered around for what they were worth. The most widely desired dream was getting in the Signal Corps and our student CO [Commanding Officer] announced to us one day that we'd have a good chance of that at the POR [Port of Replacement]. That, however, was too good to hope for, and we became adept at sorting rumor from official fact.

Passing 25 wpm had its rewards. The lieutenant passed our table one day and said to the seven of us who had by then made the grade, "My god, all you guys passed 25? Speed demons, eh? Think you could do it again?"

"Well, sir," I answered, "that's a good question. I doubt it, but once was enough!"

"That's right," he replied. "I once got 58 groups correct, but then came a group like EEASW, and I got so far behind I lost completely. It's rough!" [The Morse Code letters E, I, H and S are composed solely of *dits*. When transmitted at high speed, they go by so fast that the listener taking the transmission down in longhand can fall behind and find it impossible to keep up until the end of the transmission.]

We asked the lieutenant if he might hook us up with some code from weather or Signal Corps stations and he said he would. He did. We got WDAK, Columbus, and spent the hour listening to boogie-woogie, torch singers, and newscasts.

And then, at long last, came the closing week of the cycle at Benning. We had our final GT, upon completion of which we raised the roof. We had our last day in the code room (running our grand total up to about 425 hours of code) and, as tradition had it, lifted the rafters with a yell of joy at that last "Dismissed!" After that, the time to graduation was spent on field nets and just lazy practice. We were supposed now to be expert radiomen, ready to go out into the world on our own.

On the 21st of September, we found out just where that was to be. The company was split up and sent to a variety of locations around the country, the travel including a delay-en-route (furlough). Most of the company personnel were assigned to the 71st Infantry Division stationed right there in Benning and we felt somewhat sorry for them not to find new worlds to conquer. Others had it even worse by going to Mississippi, to the 65th Division in Camp Shelby. Some were assigned to the 42nd or Rainbow Division in Gruber, Oklahoma.

The remaining 25 of us seemed to be the luckiest. Our destiny was mapped out by being assigned to 70th Infantry Division Trailblazers, located in Fort Leonard Wood, Missouri.

There, in the heart of the incomparable Ozark Mountains, we felt we were in for a pleasant sojourn before the inevitable overseas. I was personally delighted because of the nearness of relatives in Springfield, 85 miles away. We weren't told what outfits we'd be in. That, perhaps, would be left to the Division itself. But we were satisfied. At last we knew where we were to go.

On Tuesday, the 26th of September, 1944, Class 50 graduated from its communications work at Fort Benning. This was a big day, since it meant the beginning of ten days on furlough. I managed to get most of the way home by automobile, riding with class members who had had their cars in Benning. Naturally, the ten days went like ten minutes. On the way back toward Missouri, I joined Don Masters in Amarillo, Texas, who had the same destination. We headed for the Show-Me State.

This, then, brought training officially to a close. Of course, in the service you never really finish training, for there is always something new to learn. Yet after the rigors of McClellan and the intensive studies at Benning, we felt we knew a little radio after all. It was pretty much of a pleasure to put our training to use at last.

Trailblazers. That sounded like a rough and tough outfit. If it wasn't, it was going to be. We were on our way.

FOURTEEN
LEONARD WOOD

We were pleasantly amazed at the beautiful country, the land of the Ozarks, through which we passed toward our new camp. Here in October, it seemed to have been burnt but slightly by the heat of the summer and not yet was it turning its famous autumn colors.

On the 10th of October we arrived in Fort Leonard Wood, Missouri. We pulled into the area and reported in at the classification building at Seventieth Division Headquarters. The sergeant on duty had us wait outside for fifteen minutes.

Don and I were nervous. Ants in the pants and all that. The next few moments would decide our fate. We wanted to get into a radio outfit. Sure, we'd probably get in radio, but please, oh heavens above, what about Signal Corps or some such dream as that? No, it was too good to hope for. After all, there were such outfits as infantry, artillery, engineers and all the rest. How were they going to get along without good radio operation? It wouldn't be so bad to get into one of those outfits, meet the fellows and fight the war alongside them. We resigned ourselves to it. Hell, let the softies have Signal Corps if they wanted; we'd have none of it. No, sir, us for some other outfit that fought for a living, some rough and rugged team. Wasn't that brave of us?

Then the sergeant came out and announced that we'd been assigned to the 570th Signal Company and made us the two happiest soldiers in the Army at that particular moment. Love those softies! Well, wasn't that why we'd gone through hundreds of hours of code practice?

Again in my Army career I pinched myself to see if it was real. This topped the good news; this was the thing I'd dared not even hope for. A truck took us over to the Signal Company area. We reported in there, had a long interview with the first sergeant and captain and were assigned to the radio section of the company, another blast of pleasant fortune. Logical. But who expected logic in the Army?

The captain, Morton Lefevre, was a good man. He didn't mince words. He greeted us with the phrase: "Of course, you know this outfit's hot?" He meant, of course, that the 70th was about to go overseas. Yet that was the last of our thoughts. Hot or cold, we were in the Signal Company and mighty proud to be there. It was a landmark to change the braid from blue to orange, though we knew that while we were doing this, transferring ourselves out of the infantry, we were transferring out of the finest branch of the Service. We consoled ourselves. Now, at last, the 570th had become the best damned Signal Company in the Army. Welcome Home, boys! And we were about to go overseas...

We had joined a proud outfit. The *Trailblazer Magazine* summed it up best:

"... We are the 70th Infantry, the Trailblazer Division. Our insignia with the white axe head on a red shield is the one we wear on our left sleeve. We're proud of it and of our Division. If you could see us when we're lined up in dress parade formation you'd know we're a crack outfit."

And a proud history, too:

When we first came into the Division at Camp Adair, Oregon, most of us were a bunch of rookies. We found a brand new organization. The 70th had been activated in June, 1943, and it wasn't till September that we were up to strength and organized. The nucleus was a cadre of non-commissioned officers from the famous 91st Fir Tree Division. (And that's what the tree on our insignia stands for.) Their battle cry of "Powder River, Let 'er Buck!" made history in World War I and they're in action right now in France again.

It wasn't long before we had our own traditions, though, and we'll never again hear "Oh, Susannah" without thinking of the colors going by on parade. That's our theme song, see. We adopted it because it was the song the first trailblazers sang as they opened the old Oregon Trail and arrived out there in the Cascade Mountains an even hundred years before we did. Somebody sat down one day and figured out just what the "average Trailblazer" was like. They came up with a set of statistics: five-foot-nine; weight, 149 ½; birthday, October 11, 1917; rank, not quite Pfc.; .31 married and with .13 children. (I'm still trying to figure out these last two.) Pennsylvania sent 11% of the Division, with the Midwest states close seconds. But like all statistics, these don't tell the story either. Because there really isn't such an animal as an "average" Trailblazer.

We're an infantry division, so you might expect the average soldier to be a rifleman. But in this new triangular division there are all sorts of other branches of the service represented and each man is as "average" as the next. Take the Artillery, for instance, they're an integral part of this Infantry Division and we'd have a tough time of it if their big guns weren't backing us up all the time and flattening out a path to the enemy.

Or the Quartermaster Corps who supply us with our food and equipment when we're too busy fighting to tend to those housekeeping duties. Why, this foot-slogging Infantry Division even has the Cavalry in it. Of course, they ride armored cars and Jeeps instead of horses, and they're called the Recons (short for Reconnaissance), but they're still the Cavalry, scouting out far ahead of the battle lines to seek information about the enemy and terrain we have to fight.

Or take the Medical Battalion. They don't even carry guns. But they are on the very front lines in the combat zone. There they take care of wounded buddies just as efficiently as they took care of our blistered feet and head colds back in camp....

Our good weapons are better than anything the enemy can throw back at us. The job of keeping those weapons functioning in battle—and getting them to us in the first place—belongs to Ordnance, which makes many repairs right on the battlefield and maintains elaborate repair and maintenance shops right behind the lines.

Perhaps you've got your radio on right now, listening to the news. And I'll bet you never stopped to think that our generals right in the thick of the fight are getting their war news the same way. Yep, we've got radio sets that will actually reach around the world and by this means our commanders are always in touch with the shifting of the battle. Not only radio but phones and telegraphs, signal flags and panels, and even foot messengers make a tight web of communications. All these come under Signal Corps, which is another part of the 70th.

We like to gripe about the mean old MPs like all soldiers do! But we really don't mean it, because we know that our Trailblazer MPs are a combat unit that will mop up the enemy towns we take. They stand out exposed to weather and enemy fire while they direct the traffic that brings food, ammunition and reinforcements up to us. Another combat unit are the Engineers. They build the bridges and roads the Division advances along… at the same time fighting any enemy with the audacity to challenge their progress.

And here are the boys who put the "Infantry" in 70th Infantry Division. These are the doughboys, the foot-soldiers, the guys with the rifles and mortars and machine guns and bazookas and bayonets and flame throwers. They're the guys whose job is to close in with the enemy, to take precious ground and keep it. They're the men whose home is a foxhole, who carry all their belongings on their back, who slog through mud and eat canned rations. They're the fellows who cuss and gripe and fight without benefit of headlines and who—bless 'em—win the war.

We were getting pretty well-known around Oregon as a top-flight outfit. They even named a ship for us, a huge 16,000 ton job, the largest ship built on the Pacific Coast and one that set a record for building time. The "S.S. Trailblazer" they christened her and we sent a combat team up to Portland's Swan Island yards to give her a real send off. And when the War Department set our birthday, June 15th, to celebrate National Infantry Day, it seemed as if half the state of Oregon turned out to attend the gala field day we had to mark the occasion. Shortly after that, we travelled halfway across the country to our new station at Fort Leonard Wood, Missouri, a giant military post in the heart of the Ozarks... In Missouri, too, we got a new commanding general, Major General Allison J. Barnett, a Pacific veteran who succeeded Maj. Gen. John E. Dahlquist, our first leader, who is now in southeast France... So, with good men and good leaders we have all the ingredients of a good outfit..."

Just for the record, the Division was composed of: 274th, 275th, 276th Infantry regiments, 882nd, 883rd, 884th Field Artillery, 725th Field Artillery, 370th Medics, 570th Signal Company, 270th Engineers, and 70th Recon Troop. This was a hot outfit and indeed after learning the background, we felt we had accidentally "picked" a swell group to go over with.

In the meantime, however, our own little spheres whirled with activity. We found that a whole seven of us, out of 31 Benning Class ECC50 grads assigned to the 70th, had managed to get into the 570th Signal Company. We joined them in the radio section barracks and took up our bunks. The barracks here were frame, like those at Ft. MacArthur, and they turned out to be nice and cozy, for fall was breaking upon Missouri and cold and cloudy days became more and more frequent. It was nice to be in a regular outfit, even nicer than training—for all its easy life. Here we would have no rigid training schedule or cycle, no long chow lines or endless inspections. Each morning we fell out before breakfast and spent 15 minutes exercising and running around the block. This had an excellent purpose—it waked us up.

The captain once tried out my ability at shorthand, found out how high schoolish it was, and quit. But they did learn from the forms that I could type, and as time went on I began to realize how foolish it was to own up to that fact during my first reception days. I did a lot of typing here with the 570th when I might have been out in the field doing something constructive. On the other hand, it turned out to be a good thing on days when icy wintry blasts tore around outside.

I got an upstairs bunk, along with 24 others. The boys we met were extremely pleasant to work with and we struck up friendships in no time flat—friendships we knew must last through combat. We'd gripe and moan and enjoy things together. One of the things we moaned most about was being awakened in the morning. If this process had been accomplished by chimes or radio or even bugle, it might not have been so bad. But it wasn't that way. The CQ always seemed to have the best lungs in the outfit and he would utilize them first in blowing a policeman's whistle that would blast us out of bed and then yell with a voice that brought us back up on our feet. No one could sleep after this, but if anyone fell back to sleep, pity them when Sergeants Lister and Morton came up and dumped them out on the floor. A few jolts like that brought the recalcitrant lazybones up with the rest of us.

"GET UP!!!!"

Each morning, we rose, grunting and groaning like old hermits with rheumatism. Maybe we'd stumble downstairs to wash the face and comb the hair and other usual morning functions, but it would almost never be finished in time. The whistle blew outside and First-Sergeant Pepper's foghorn ***"FALL OUT!"*** found us racing, half-awake, upstairs, throwing our towel and comb on the bunk and flying back downstairs again. Then the running-jumping exercises, then chow, wash up, clean up, make the beds, sweep and mop. Maybe even a couple minutes of bull session before falling out again. That's how we started out days with the 570th at Leonard Wood.

Being a Signal Company, they were already full of ratings. Corporals, sergeants and up. T/5s and T/4s ran all over the place. So we forgot about getting ratings… as if we wanted them anyhow. But Lister told us maybe soon…

Some training began, but this was adapted to the overseas routine and the working together of the radio section as a whole unit. We had code practice and nets, jamming and interference, procedure and field work. In between was other work slightly off the technical side. We fired on the range, and we helped others fire. I was a range guard one day, stationed all day with nothing but trees, birds and bees for company, way off down a lonesome road at the entrance to the offensive sniper woods range. We had to pull targets for other companies to fire, even on Sunday. We were issued caliber .30 Carbines (thank the Lord) and fired them on the range. We also passed through beautiful country going out to the range where we fired the M-3 submachine gun.

From time to time, we heard how our Benning cohorts were faring. Not so well. Bob Hill became a wire man in Artillery. Henley made artillery, too, said he got a little radio but mostly did odd jobs around the company. These and others we met commented on our extreme luck in landing Signal Company and invariably wanted to know if there were any openings left.

On October 25th we started drivers' school. Most of us rookies from Benning were assigned to radio teams as drivers. This we liked. It seemed that each radio set was completely mobile. After lugging SCR-284 sets on our backs, banging them in and out of ravines in Talledega and at Benning, nothing made us happier than to see our sets mounted in vehicles. Each radio vehicle had three men to a team under combat conditions. We were thus assigned now, though not on any permanent basis. And we became drivers. This involved learning the techniques of driving a number of company vehicles. We went out on caravan through the rolling Ozark hills, winding down country lanes, across creeks, up cliffs—so that we could have a good time stalling the truck—and even out on Highway 66. Our radios were mounted in $^{3}/_{4}$-ton weapons carriers, which meant that much of our practice was with these. We drove other types, though, and the ones I much preferred were the $2^{1}/_{2}$ and 6 ton trucks because there was ample room for my knees.

Such was not the case with the Jeep. My knees would come up and wrap around the steering wheel, and I would have to decide each time whether to shift gears under my leg or over it. Once I moved my knee in order to shift while going up a hill, hit the ignition key and turned off the motor. Getting down to the accessory shift involved plunging down through a maze of legs, knees and gears. But I made it and actually began to like the Jeep, especially with an eye toward civilian life. It was capable of traversing the most rugged terrain and seemed perfect for exploring the last road beyond the last mesa in red rock country. At length, we were issued our driver's licenses and became full fledged drivers.

However, the glamour of possessing our own trucks lessened by degrees. Literally. At first, getting up and servicing them in the morning wasn't bad. It was done before eight: start the engines, warm up, check gas, oil, tires, radios and other odds and ends that consumed perhaps half an hour. But as summer left and fall took over, the mornings grew chilly. Then with the first hint of winter, the mornings became icy. So we shivered in the cold as we serviced the vehicles and wondered why we ever had to become mechanical experts.

After we graduated from Benning, we had become medium speed operators, MOS 740. Now the next goal was to be MOS 766: 20 words per minute pencil, 25 typewriter. We practiced from time to time in the code room at the company area and got some practice sending. I tried using the "bug," a high-speed sending instrument that would enable anyone but me to send at speeds around 40 wpm.

Our radio sets were SCR-193s and some were mounted in carryalls. They were big and it took us several days to learn their complete operation and primary repair. Then we went out in Division nets, parking the truck in some convenient place, rolling up our sleeves, sitting back and taking life easy. At other times, we operated mobile, almost always as Net Control Station. These were good times. We learned a little about the company's biggest set, the SCR-399, which occupied the entire back end of a 2½ ton truck and required an auxiliary trailer to house the generator and extra cans of gasoline. We got a little practice, too, with teletype and with switchboards, here 44-drop affairs much larger than those with which we had become acquainted at Benning.

Lieutenant General Ben Lear and his staff visited Leonard Wood while we were there, presumably to approve its overseas departure, we weren't sure. Included on his staff was a little fat Signal Corps colonel who found some dust on radio tubes in the control room and the entire radio building in a sad state. He had a high time questioning the officers and NCOs of the company, getting them well muddled. They took out their embarrassment on us by having us clean up the radio school.

We had a basic signal class in which we learned the functions, equipment and work of all the sections in the company. Some of it was repetitious of Benning, the rest interesting. We were shown a movie covering defense against jamming and had some practical experience along this line. Not only did we take several hours of code mixed with jamming and/or interference, they installed a loudspeaker in the room and put on a recording of battle noise: the mingled roar of planes, bombs, machine guns, rifles and varied explosions. All to keep us from getting addled under fire. This and other training was designed to make us combat serviceable.

In the middle of November, the lid blew off the icebox and we turned into ice cubes. That little trick of exercising 15 minutes before breakfast (now before the first rays of dawn) grew colder than the North Pole climatic zone. We had a dress parade and were merely automatic walking icicles; we couldn't wear gloves, of course, and you never saw so many blue hands and faces. Four of us, apparently thinking we loved the cold, hopped in a Jeep one day and rode out in the countryside. The Jeep was sans top and sans windshield. When we got back, we were sans movement. About this time, all the drivers drew nice warm driver's gloves.

I won a steak supper once and we went into Waynesville one evening so the loser could pay up. The steak was good but we at first doubted if anything could be good in that town. Even before the war, Waynesville (six miles from camp) was probably little more than a wide place on Highway 66. Now it had growing pains because with the opening of Leonard Wood it was a boom town, and the Army had ruined it. Since drinking was frowned upon officially in camp, GIs would use Waynesville as an excuse to get away from the Post and get drunk openly. This, it seemed, Waynesville couldn't take, so it surrendered to the inevitable horde and allowed the lost weekends to turn it into a shambles. Streets, stores, the whole town presented the one appearance of being dirty. I began to think of this process as typical of small towns next to GI camps.

Anniston in Alabama and Columbus in Georgia would have received the same treatment had they not been bigger. They had a place to shove the GIs who wanted to get drunk; Waynesville did not, so another charming Missouri town bit the dust.

Rolla was another, larger place farther down the road, east on Route 66. We were to have a party there once because the company had $600 in its fund that it didn't know what to do with. A beer party, by popular demand, with fried chicken, all the trimmings, and dancing on the side. At a company meeting the officers told the men that they were to go and have a good time and to drink all the whiskey they wanted, the more the better. With that high brass sanction, some of us expected the thing to turn out as a filthy drunken brawl, which is exactly what happened. Half the company didn't even go. There are better ways of being dragged out from under a table. But then you had to give the kids credit. After all, they were going off to fight and die for their country.

"Parties are so gay, don't you think?"

The impressions of Fort Leonard Wood, disregarding beer parties, stand out now as pleasant to a great extent. It was a good camp, the weather was never dull. Fall brought clouds and cold and rain, but I think that did me good. I'd been getting too soft on the climates of Georgia and late spring in Alabama.

Theatres and PXs and bowling alleys and the field house were always there for our entertainment and exercise. Always in the background while we were there was the fascinating panorama of the Ozarks turning from summer to winter.

Unquestionably the greatest impressions left from the days at Leonard Wood were the weekends in Springfield. The pass book was open most weekends we were there, and some of us would crowd on a westbound bus and travel to the big city. I had the advantage because of the big passel of relatives there, headed by Aunt Florence. The times we spent there were times in Heaven, not only from the outstandingly pleasant atmosphere, the kind hearts of all who came there, the trips and the games we played, but from the meals to which we were treated. I took a friend or two each time, and we reveled in feasts of roast hen or turkey, and delicate meat loaf, with all the possible trimmings. Aunt Florence renewed her fame each time in making lemon pies. It was seldom that we ate anything on Monday back at camp. I felt that I had come so close to bursting that nothing in combat could be so hazardous.

Aunt Florence's classic statement was this: "Whatever I do is little enough for what you grand boys are doing for us." And she did plenty. She worked almost seven days a week at O'Reilly General Hospital in Springfield, spreading her kindness among the sick and wounded there. During the week she had patients to her house for supper, then us on the weekends. It's no little task to prepare a meal for hungry GIs who haven't tasted such cooking in months. One morning, she got up at five a.m. to bake some pies for us before going to work. Aunt Florence's kind are just too rare.

An extra bonus: while there, we enjoyed the privilege of waking up slowly, something impossible at camp. We began to wish to be stationed at Leonard Wood all our Army career. Or perhaps at Camp Crowder, a Signal Corps school even closer to Springfield. At any rate it was a lovely way to spend a weekend. The temper of many civilians in those days was to do all they possibly could for "our men in uniform."

Such were the first days at Leonard Wood. We were at last becoming vital parts in a giant machine that would someday win the war. This was our outfit, and we liked it, liked having a part in making it what it was. And yet, as October drifted into November in 1944, we couldn't help but notice what went on all around us, out beyond camp. Fall came to the Ozarks, one of the most famous of American mountain ranges at all seasons.

FIFTEEN
OZARKS

The land into which we came at Leonard Wood intrigued us immensely. With always something new to see, some new twist of nature's imagination, we watched as each bright living spectacle shot across the countryside. When we went into the field, the fragrance of burning leaves assailed us, indicative of life going on normally in an abnormal world at war. We saw pumpkins, corn in the shock, leaves raked in piles, the people making ready for winter.

During driver's school, we journeyed through golden and scarlet landscapes on some of the most decrepit semblances of roads I'd ever seen. That's saying a lot, considering my years of traveling western byways. Occasionally, we took off cross-country on no roads at all, merely following two tracks of beaten-down grass. If someone had gone there before, so could we—in theory. Up and down hills, perilously and probably stupidly, taking our lives in our hands.

We came to a rickety bridge once after getting lost on the way back to camp from the rifle range. It was a bridge across the Big Piney River and I dared not drive across with anyone else in the truck. I stopped. We got out, walked on the bridge (swayed would be more descriptive) and began to meditate. The green waters below, resplendent in the afternoon, seemed so deep that the truck would almost surely vanish from sight if it fell in. Getting lost was already one blotch against my reputation. Losing a truck would be a real demerit. If that was what I was going to do overseas with a massive truck full of expensive radio equipment, they would be better off tethering me on the island of Guam.

There was simply no way we could tell if that bridge would hold. I imagined that not even a construction engineer could tell without weighing the truck and somehow testing the aging bridge supports. There was no going back. We were already so badly lost that we'd probably run into another series of bridges just as bad, cross one, cause it to fall behind us, and get trapped between two such bridges. The engineers would have to construct a new bridge to get us home. Think of that humiliation! My imagination became very fertile at times like this.

An oldtimer came along and we eventually gave him to understand our difficulties, but he couldn't predict either if our 2½ ton mechanized giant would make it safely. The bridge had obviously been designed for horse and buggy, at most a Model T.

Well, we couldn't stay there all day. So I climbed in and drove forward. The front wheels bumped up onto the boards, which creaked (or, as I thought, cracked). I put my head out the window, face white, bloodless knuckles gripping the steering wheel. I watched the boards sag under the wheels as the full weight of the vehicle passed onto the bridge. I held the door ajar so that I, at least, could leap out if I felt the world falling out from under me. Inch by inch, I listened to the grunt and groan of the straining boards and girders. Half way across. Three quarters…

I made it. We all collapsed with immense relief. And not only that, we found Highway 66 an hour later and got back to camp before chow. I was the hero of the day. The perils of war.

Still, those winding roads led us through exquisite valleys, rolling farm country and along the Osage, Gasconade and Big and Little Piney Rivers. We passed maple, hickory, oak and walnut trees. Having a taste for black walnuts, we looked for some but found none. Inquiry at a farmhouse brought out the information that this year had been a bad one. Although much of the terrain constituted Mark Twain National Forest, we passed rustic farmhouses now and then, some set back in coves, some on the road. Each boasted a horde of pigs, chickens, women and children. In many places, the Ozark region still retained vestiges of its original wilderness charm. Here and there, sandstone cliffs tower over streams, meadows, grassy vales and pasture lands.

The Big Piney River drew us back like kids to an ol' swimmin' hole. The river gathers in languid pools, flanked by soft beaches, moss-covered banks, and grassy meadows. Sycamores held their leafy branches out over the water. The whole company went out on a sunny October afternoon, and we swung on ropes far out over the water, sailed through the air, crashed into the pool, and sought to see who could swim farthest under water.

No cry of peanut vendors here, no clank of milk cans, or mournful howl of ferry boats. The Ozarks stood out like an island in a sea of noise and frenzy. My regret was not getting to know the country people better. Not getting to their spelling bees, corn huskings, hog killings, and square dances:

Lady go gee, gent go haw,
Around and around in a promenade all.
Swing your taw,
Across the river to yer ol' Granmaw,
Now to the girl from Arkansaw,
Swing around to yer old Granpaw,
Now your taw and promenade all.

We met some of the people, and reveled in their stories. They told of fields so steep that a mule fell out of a cornfield. They told of the pumpkin that grew on the hillside, broke loose from its vine and started to roll down hill: "Hit bounced once't and flew clean over the house, hit agin in the corner pasture, and last I saw it was goin' right on over Turkey Hill!"

Some times a seasoned farmer might be heard to say: "Some times I sets and thinks. And sometimes I jes' sets." Ethel de Long, in *The Far Side of Pine Mountain*, quotes a mountaineer as saying: "We-uns that cain't read er write have a heap of time to think, and that's the reason we know more'n you-all." The noted geographer, J. Russell Smith, says: "If the time ever comes when the requirements for citizenship are based on intelligence rather than information, perhaps these people will make a better showing than the multitude in cities who have just enough education to read cheap dailies."

Cecil J. Sharp wrote: "I'd like to build a wall about these mountains and let the mountain people alone. The only distinctive culture in America is here. The people *live.* They sustain themselves on the meanest food. They are not interested in eating but they have time to sing ballads."

A missionary was once taken to see an old bedridden man. The missionary's companions told the old man of the progress a visiting lady was making at learning to card and spin wool. "I didn't s'pose," he replied, "ther was ary woman in the whole world didn't know how to kyard 'n' spin."

Such was the atmosphere that surrounded us at Fort Leonard Wood, a culture we touched so lightly. We would have liked to stay longer and see more, but that wouldn't quite be possible. The Army had other plans and who were we to argue?

SIXTEEN
WINTER

Our departure from Fort Leonard Wood sort of gradually grew on us. It was one of those things we were naturally expecting, knew it would come, and yet it was filled with that uncertainty as to when? When? When? Rumors flew like birds on migration. Some of the wild stories we believed because we wanted to believe them. Some we discounted. After a while we forgot them all and the first person who mentioned a rumor in the barracks was severely ostracized and practically thrown out on his ear.

Still, the rumor fever in those November days caught hold of the officers and the NCOs who knew as little as we did but tried to tell us something that had the stamp of officialdom on it, probably to soothe our tired souls. At formations and company meetings, our CO (Commanding Officer), affectionately known as "Big Mort," would impress us with the fact that there wasn't much time left. As if that was any consolation.

So we began slowly to get ready to leave, to cross the blue horizon and fight for our country. They told us to get rid of all our personal equipment and not attempt to take anything to the POE (Port of Embarkation) that might be taken from us. So all those precious little gems and souvenirs that connected us with home were sent right back home along with tears of sadness. This also applied to men with wives, since the wives would undoubtedly be taken away from the boys at the Port. The authorities advised all married men to send the little woman home, and pronto. Most of them did, before the last moment. They made their sad adieus, then sat and moped in the barracks for a while, then pitched in and helped the rest of us get ready for the big move.

We got lectures on Military Security and Censorship. There had heretofore been much speculation on which ocean we would cross, which enemy we were to fight. I seemed to be the only one in the barracks who had a bent toward the South Seas; the others all preferred Europe. I suspect my wishes were derived from visions of swaying South Sea hula-lovelies whose images I had projected in the theater in an endless stream of film dramas from *The Hurricane* to *Song of the Islands.* I dreamed of palm fronds waving in the breeze, bananas stretching in bundles overhead—to be had for the reaching—and emerald coral seas for swimming. Of course, war in the Pacific wasn't like this rosy picture altogether, but I was too stubborn to be convinced. I still wanted to go west, young man. But I soon gave this up, for we began to get strong hints as to our secret destination.

We had classes on the German Army. We had brief language sessions in French and German. We were issued warm wool clothing. This was enough. We were convinced. I changed my hula-girl vision to a solid mademoiselle lounging in one of those ooh-la-la French swimsuits on the sunny Riviera. Wasn't the Riviera always sunny?

So we began to dream of swimming in the Mediterranean. Waltzing in Vienna. Scaling the Matterhorn. Cycling through Paris. Boating on the Rhine. Freezing half to death on the line in Germany.

During the last days of October, it seemed that every day new guys were coming into the various units of the Division... other Special Troops outfits, the Infantry regiments and even into our own company. We felt then that overseas was a long way off. When it came that would mean getting over to a staging area far from the combat zones for several months of maneuvers to get acclimated. Certainly the Division was too raw to go over as it was.

It turned out that this was decidedly a sack of nonsense. Not only was the outfit hurriedly and efficiently equipped and trained through November, but the idea of loafing around overseas before hitting the line was pish-posh.

Twice a week, we began having dress parades. Soon fall disappeared and the icy blasts of cold weather came full upon us. Then came the first light snowfall, just to convince us that our days of enjoying summer weather were over. We chose every possible means of staying inside on days like these. I was trapped in the supply room typing, though sometimes I would have preferred an outdoor job, ice, snow and all. I was typing the packing lists that were sent along with our equipment and the lists were addressed to the Boston Port of Embarkation. Of course, they could ship the stuff through the Canal and over to Hawaii. But then I gave up the last glimmer of hope, took on an up-east accent and practiced a little more on French and German.

On November 15th, the first movement began from Leonard Wood—one of the Infantry regiments pulled out. The other regiments followed shortly thereafter and only Special Troops was left. We still didn't know when it would be our turn. At least there was a period on into December when life at Leonard Wood was free and easy. With only the few troops left, that meant no lines at PXs or theatres, no more crowded busses into Springfield, a generally all-but-deserted camp.

About this time, I heard from a former Hut 9 pal, Chuck Griggs. He'd already been through Belgium and France and was at the time in Germany. Said he was in an Infantry Headquarters Company as a radio man and wasn't doing at all bad.

Twelve of the best men in the Radio Section were relegated to leave Fort Leonard Wood along with the advance Infantry units to make things ready for the Signal Company when it arrived at the POE. They left in the middle of November. The Division Signal Officer left with them, but before departure he talked to us. We would receive full information, he said, on what to expect, where we were going and when, plus all the official G-2 we already knew. His little informal chat turned out as so many others had when big news was in the offing. We were left exactly where we started. He said: "Goodbye until we meet again.

We'll get a little better acquainted over on the other side." And they were gone.

Even to this time, some of the most optimistic souls held to the suspicion that the way the war news was going, the 70th wouldn't even get into battle. Takes a long time, they said, to move a whole division overseas and certainly by that time the Allies will have whipped the Germans. Nevertheless, we diligently went ahead getting ready for the trip, and gradually becoming combat serviceable.

A new trouble developed. I had to smooth over all these activities in those letters home. Like all Mothers, mine could see me getting killed in all sorts of horrible combat ways, and I had to play up how safe it was going to be. I'm going off to war, and it's going to be safe? What intelligent mother would believe *that*? I drifted off into vague comments about my activities, even a few little white lies, and concluded by saying "My address is changing. Write me at the APO (Army Post Office)."

On Thanksgiving Day I had KP at officer's mess, and offhand I can't think of a better day or a better place to perform that duty. By the day's end I was stuffed tighter than the turkey had been.

Then the weather turned frigid, and the mercury began a nightly dip to zero. This brought out our heavy GI overcoats which, if walked in more than two blocks, began to bear down on the shoulders like a sack of potatoes. No coat was long enough for me, so I felt like I was trussed in a waistcoat. My legs got cold. I tried to wrap the flailing wool around my legs, but then I couldn't walk. The solution was to get back indoors as soon as possible.

The trips into Springfield continued up to the last. One night, Uncle John took us to jail. He was a detective on the Springfield police force, and took us straight to Cell #3, whose history had been written deep in blood. Turned out that three men had committed suicide there, two by hanging, the third by the quaint process of cutting his wrists and throat with a tin can lid. He told us the vivid, bloodcurdling story of how six policemen died in a gun-battle with the Youngs brothers way back in '32. Hardened to combat as we were, his tales of death and transfiguration were right down our alley.

While in Springfield, we decided that we'd better get our Christmas over while the getting was good. So I got all my presents to mail, and Aunt Florence insisted on wrapping them. Then I bought presents for all the Springfield families and delivered them in person with curt instructions about not opening until Christmas. On our last weekend in town, we had a Christmas party, decorated the tree, and had all the thrills of the Yuletide, even though we were three weeks early. This didn't seem to bother us a bit.

It was announced that at the POE we would be issued a turtle-neck sweater, wool, and six pairs of high-top ski socks. They were making it apparent that we were to be assigned to some cold combat area, say, Greenland, Iceland or Norway—just any old igloo would do. On the other hand, this information might have been for the purpose of misleading us, though where that thought originated I'll never know. The theory was advanced that if we leave a New York POE, as was indicated by word from the advance party, we might make a U-turn and wind up heading the other way. I didn't advance it, honest… I was limp by now and not giving a tinker's damn where or when we went. When we began to think we'd never leave, December came to the land the Ozarks and the excitement started.

SEVENTEEN

1508-E

On the sixth of December we went out to the cold fog-covered ranges at daylight to fire our sub-machine guns again. On returning, I spent two hours preparing my clothes and equipment to be turned in, saving only that which was classified as "Combat Serviceable"—new stuff that would stand the gaff for several months if need be. I turned the stuff in after dinner and drew my watch. This was a pleasure, and, too, a business necessity, because radio operators must always have the time correctly on hand. For that matter, most of the company drew watches. Mine was a Waltham, and a day's checking found it to be perfectly accurate.

On the eighth of December, we turned in our foot-lockers and began to live out of barracks bags. As the days wore on we found ourselves waiting. We went to the radio school and took some code or press reports or practiced sending. We'd write a letter home or go see a movie on camouflage or some similar combat subject. We attended French and German classes and spent a lot of time simply trying to keep warm. Snow in icy gusts whipped about the corners of the buildings and laid down an icy covering on all the walks and streets in the company area. At last, though, the packing list typing ended, and I relaxed to wait with the rest of the boys.

On the 13th of December, after ten months and eleven days in the Army, a special 570th Signal Company order came out promoting certain members of the radio section to the rank of T/5. Now it was Corporal Sut and I learned a lot of sewing techniques that afternoon.

Stripes at last! I went around the Post that evening pulling my rank on any privates I could find. We were just glorified Pfcs but it felt good just the same to slap "Cpl." on that return address. Vanity is the same everywhere.

We had a big radio test one evening, covering everything we were supposed to know. The following evening from six to nine was spent in answering and discussing questions. We had a complete check of all our equipment to make sure that we weren't carrying anything to the POE that we weren't supposed to carry. And speaking of POE, Orgie had written Stinky (these were the actual names conferred on some of our confréres) to say the advance party was on the East Coast somewhere and that the barracks weren't much to talk about. Of course, the Infantry and Headquarters boys were now already across and almost in combat, but it was a sure bet that POE conditions hadn't improved. And snow had set in, so apparently they were north of the Mason-Dixon Line. I shivered. I was never very well made for winter warfare. I knew that before the whole thing ever began.

All the trucks were turned in to Ordnance. The Radio Building was removed of all equipment and we cleaned the building thoroughly. Our French and German classes were brought to an end, and we found that we could get along fairly well in a strange language. I'd always had a soft touch for languages and was intensely interested in these preliminary classes. But then, I figured it wouldn't be long before I had a cute little petite in France who could teach me the language so much faster. Beyond that, I wondered if they were going to teach us any Chinese, merely because that's a language I very much wanted to learn.

About the hardest workers during this period were the men on the Packing & Crating detail. Their job began a long time before we left. It seemed that our company had the most equipment of any other outfit in the Division—Signal this and Signal that, and all the odds and ends. Much of the stuff on hand was usable, so it was packed, waterproofed, stamped, labeled and recorded what seemed a hundred times. During the early part of it, I was in a swirl of papers while typing the packing lists. When everything was finished once, someone else wanted another batch of carbon copies so it all started over again. There were about 250 boxes and some contained as many separate items, with serial numbers to boot. Times like that you find out what it takes to move a division overseas. The buildings were GI'd to the Nth degree while the nights were swept with a fierce wind filled with snow.

On the 15th of December, we turned in our sheets, pillow case, mattress cover, a blanket and two comforters. And what a ridiculous time of year to do this type of thing. The next few nights found not only snoring in the barracks but the violent chattering of teeth.

On the morning of the 16th, the General talked to us. Like all good generals, he told us that we were a fine bunch, and that he hoped division personnel would do as well in the difficult times ahead as they had done here and at Camp Adair. And all that sort of pat-on-the-back I'm-expecting-great-things-of-you stuff. We liked him. I guess that's because we never did see much of him.

Nothing is ever handed down to the troops as a certainty when it concerns a vast movement as this one was, and we were never sure of the day we were going to leave until the night before. We had a physical inspection, a sham affair officially titled POM, Preparation for Overseas Movement. It was only supposed to be in preparation for a final, complete and thorough exam to be given at the Port, where it would be determined whether or not we ourselves were CS—Combat Serviceable. There were rumors about that Sgt. Pepper, more intimately known as Skidoo in our lingo, wasn't going to get to go over with us on the mere fact that he wasn't popular with the men of his command. That was cleared up one morning at Reveille. He told us he was disgustingly healthy.

We had a final showdown inspection in the Medic's Rec Hall and it was at this point that we were the most completely GI we'd ever been. On the evening of December 18th, I wrote the following letter home:

> *I've spent my time reviewing French & German and doing my last minute packing. Tonight I shall go to the show—and then get to bed as early as possible. Tomorrow I'm on KP and I feel that it is going to be an extremely rough day."* (KP on troop trains usually is.) *"This may be the last letter you'll get for some time now—I don't know how long till the next one will come through.*
>
> *It will depend on the authorities at the Port. But don't worry—I'll write every chance I get.... My next letter will*

probably be censored..." (It was.) "Just remember, everything is going along fine and your kid the corporal is ok, so don't worry.... Our mail has already been forwarded...Hope to get it at the POE."

The last night in Fort Leonard Wood was as cold as a frosted doorway to an Eskimo icebox. And us with one blanket. As I lay there telling myself those weren't icicles down my back, I wondered what this POE was like and what was going to happen there. I wondered where it was.

They'd painted a not too rosy picture for us and our own imagination had been a good canvas for the oils that went into this picture. The General had told the Commanding Officers to tell their men not to expect passes when we got there, but that there might be a few issued to us if he was lucky in pulling somebody's strings. They said there were a lot of things that we couldn't get at the Port and to stock up here at Wood, so long as it didn't interfere too overly much with our keeping strictly GI. We were informed that what we couldn't get before leaving the US we might just as well forget about.

I wondered if I'd be able to get away with keeping my all-protecting wool scarf and those extra handkerchiefs that I had tucked up the sleeves of my wool undies deep down in that duffel bag. Skidoo had verified a couple rumors by describing the heavy wool clothing we would be issued at the Port. I shivered a little at this.

Then I went to sleep and the icicles turned into overseas stripes. Next day I got up early and grumbled all the way through dressing. Of all the days to be on KP this was probably one of the most undesirable and I think I would have enjoyed that day a little more the other way around. It was December 19th.

We ate and served breakfast before dawn, then loaded the last remaining kitchen items into a 2½-ton truck and rode down to the rail yards as dawn came. The kitchen car had been cleaned out the day before so we moved right in and started setting up for a dinner on the road to our first secret destination.

Next came the trucks loaded with duffel bags, that latter stored in the adjoining baggage car. Shortly after this, more trucks poured in and emptied out the company, followed by other trucks with the rest of Special Troops. Watching from my kitchen car ringside window, draped over a potato sack, I saw them line up outside the train and receive last minute instructions. Like I said, I would rather have been with them than with the potatoes I was neglecting at the moment. Without much ado the various sections filed into the waiting cars—Radio in one, Message Center in another, Construction in the third, and so on down the line, each section with a car to itself. From then on, no one was permitted to leave that train at all during the trip and moreover to engage in no conversation whatsoever with anyone on the outside while we were en route. It would be considered a court-martial offense to attempt to mail a letter of any sort by throwing it out the window or giving it to a civilian.

We were now a secret outfit in a secret movement. We had torn off our shoulder patches and sent them home. We ceased to mention our division in any letter and had henceforth to destroy all incoming letters after reading them. In effect, the 570th Signal Company, 70th Infantry Division, was wiped out. We became merely a number and from now on would be known only as Unit 1508-E, pronounced Fifteen Oh Eight Easy. With this, the strictest military censorship came into effect.

At 10 a.m. unit 1508-E rode out of Ft. Leonard Wood, Missouri. This, brother, was it.

P.S. Remember! The enemy is listening…

EIGHTEEN
MYLES STANDISH

A few hours after leaving Aunt Florence and the ASFTC behind, and by the time dinner was served in the kitchen car, I knew we were heading for the east coast—and points beyond. I was beginning to feel pretty sure where those points were going to be.

Meals on a moving troop train are rather an ordeal for those unaccustomed to such things—as like sea travel. Thus it was highly amusing to watch the boys come in one end and go out the other, picking up on the way a paper plate, fork and cup. The line reversed then, and they picked up their vittles on the way back through. And from where I sat dishing out bread and butter, it was quite a sight. With every jolt of the kitchen somebody flew off balance and went sailing off down to the door, not favoring a return trip for their bread or whatever had been missed. A few embarrassing scenes occurred and Norman the cook swept valiantly in to clear the path.

Going through the cars was fun, too. After so much coffee had been spilled on the floor, it became common practice to slide into someone's beef stew. A few plates of food went sailing past a few ears, and a few cuss words got lost in the mêlée. Eventually things turned out OK with a little grumbling, though we lost a lot of good meals that way. Seriously, though, the meals were pretty good on the train, considering the conditions under which they were slammed together. That's not much of an overstatement. Maybe it was the country passing by the window that took our minds off the food, I don't know, but it was at least different from the mess hall we'd gotten tired of.

The country we passed through was beautiful enough, if you liked the Missouri Ozarks, and I did. I spent much of that first afternoon looking out the kitchen window and then looking at myself in retrospect and asking "How in the hell did you ever get yourself into a place like this?" But it was a good ride. Just before supper we hit St. Louis and stopped in the yards for a while, then proceeded on up the Mississippi for a short distance before crossing it. I practically had my head out the window going over that bridge. It seemed to be the gates to home that I was going through, but I kept saying to myself that it wasn't as bad as all that. We'd be back soon. Just kept kidding myself along.

As with all KPs, this one ended soon after supper and I went into our car and sat my weary bones down with Dody Ratigan and Luke Potuzak to enjoy the darkening scenery from a comfortable position for a change.

On eastward in the snow and sleet of the night, past exquisite moonlit snowfields, patches of light, cities. Night through Chicago and across Indiana, then morning in Ohio. At Kent, we got off the train to stretch our legs and do a little double-timing. Across a corner of Pennsylvania and into New York. Here the trees had all lost their leaves and were covered with a light snow layer. Most of the day was cloudy. Then Syracuse, Schenectady, out of New York and through the endless Hoosac Tunnel—purportedly the longest railway tunnel in the world. And night came.

We couldn't tell much about Massachusetts and even less about the camp at which we arrived shortly after entering the state. The call came to disembark. We hoisted our overcoats, shouldered our duffel bags and got off the train. We found ourselves in nothing more romantic than an unloading platform whose darkness was broken only by a few pinpoints of light down the track in front of us.

The engine sent out a few wisps of steam. All up and down the line, other GIs climbed off and harkened to the shrill calls of ***Fall in!*** and ***Attent-hut!*** Officers scurried about, looking important and seeming to appear in a great rush. I watched them when our section was rounded up and wondered if there was really that much fuss and bother and that much to get excited about. Why the hurry? The name of the camp was whispered down the line and finally reached us. We had arrived in Camp Myles Standish, Massachusetts. It was December 21st.

We got to our barracks and found them to be exceptionally long, with double-tiered bunks down each side and the middle. There were two stoves, one at either end. It didn't take us long to get these started, to toss our bags in a corner and hit the sack.

"I see you've been fitted with a new mouth."

Next day, we had an orientation on how and how not to write home. Suggested writing topics were: your friends, bull sessions, non-specific battle stories, descriptions of the interior of your billet or foxhole, customs of the country, local farming methods, clothes the natives wear, movies, clippings (without remarks), YANK feature stories, letters from friends, general quality of the weather, and love. We were given a long list of taboo items, but I merely summarized the most important: (1) No details of movement, to or from the European Theatre of Operations; (3) no strength, efficiency, training or morale of any outfit; (4-5) no locations whatsoever; (11) no cracks about Allied armies; (16) no games, puzzles, maps or blank paper. This indeed left very little to write home about. As a matter of fact, my first censored letter went like this:

I imagine it will please you greatly to hear that I arrived safe and sound—in good spirits and in good health—at our secret destination, which is, shall we say, "somewhere (deleted by censor). *As is my usual custom upon arriving at some new place, I proceed to tell you about it... This time, I can only tell you that the food is good, the weather is cold... the camp is fair and the barracks are so-so. We are located near a theatre, service club and PX—and means of recreation here are many and varied. The barracks are warm—we have fire orderlies on 24 hours a day. There are lots of details, such as KP and guard duty. This morning we went to an orientation and found out the various things about writing under censorship (telephoning and telegraph, too), conduct, when we get passes, and things we must know in order to keep our movement, our unit, everything in secret...*

Censorship was strict. When I finally got around to reading the bulletin board I had to sit down and apologize to my correspondents for having written a letter that came to them in shreds, and to clarify things that were snipped out. As a matter of fact, one letter came back so badly chopped up that I was advised to rewrite it in less specific terms. One thing threw me off. The boys in the advance party had written from the POE and stated they were "Somewhere on the East Coast," so I put that immediately in my first letters and it was just as immediately chopped out. It seemed to be more profitable to say "Somewhere in the Eastern U.S." Though I sometimes wondered if they weren't just a little too scared about the whole thing, I finally resigned myself and did what I was told. And that's one way of learning how to write a restricted letter: classification confidential. Try as I may, I never could get any information to the enemy.

The 2½ weeks at POE weren't characterized by any major events as far as barracks life was concerned. We didn't see too much of our barracks because we were out to raise as much hell as possible before leaving the country forever. Occasionally we dropped into the barracks to sleep or write a letter, and time was spent in bull sessions around one of the stoves—sessions that included every conceivable subject that could possibly be discussed around a stove. Mostly, though, it was about the war and its outcome and our place in it. The boys began to think a little now and came slowly to the realization that this business of going overseas was somehow becoming real. So they started to forget about the scrumptious babe in Manhattan and concentrate on last minute details before my bonnie was to lie over the ocean.

We had our physical exams and received the last pieces of equipment. Our combat boots—here was a dark cloud in the overseas sky. As if my duffel bag wasn't already full to the hilt, I was issued two pairs of size 13s rather than the customary one extra pair everybody else got. I griped to beat the devil, but they warded off the attack by saying mine was an unusual size that might not be so easily obtained overseas, and that I'd better take along this extra, extra pair so I wouldn't be fighting the war barefoot. I felt that by the time I wore out my current pair of combat boots, they could send over a whole shipload of them, but I swallowed my epithets and stuffed the shoes into the duffel bag—after which I felt no pity for sardines in a can. How I was ever going to hoist that bag up the gangplank, I didn't know, but then there would be extra tugboats and cranes in the harbor.

Though the work details were numerous, they were seldom difficult. Being latrine orderly (or would you call it Captain of the Head?) had its compensations in that it gave us a vast amount of time to shower and clean up without hurrying, to wash and scrub clothes, oil boots, read, write letters and sundry other tasks that become last-minute when one is departing for a voyage abroad. There was spare time around the barracks, especially through the Christmas holidays and I managed to catch up on a lot of reading I'd been wanting to do for a long time. Something cultural, to improve my intellect, like *Ranch Romances.*

Massachusetts can be very cold in winter. There can be snow and there can be ice. We had both. Together. One of the funniest things we had was the process of trying to march in dignified cadence down streets that were covered with a coating of ice. You can imagine this if you can visualize marching straight-backed, overcoat and all, across a smooth, ice-covered pond. That's what it amounted to. At first the sergeants would try to count cadence so that we could keep in step, but they gradually gave that up and though we never were officially given rout step, that's what we had. I would go along proudly in step, boasting inwardly of being so sure-footed when all of a sudden one foot would slip out from under me and straighten out with a crack that sounded like the leg bone had been fractured clear down to the ankle. I'd stagger around until the other foot lost traction and have a severe time of it until I got back in position. I never fell all the way, though, but came so close it wasn't funny. Some, however, did. They lost control of both feet at once and crashed to the ice with a terrific clatter. This would sometimes set up a domino chain reaction and knock the whole squad down. I'm glad they didn't allow civilians onto the Post. It would have been too mortifying for us combat soldiers. At that, I began to wonder when our ice skates would be issued.

With this weather and climate, I naturally got a cold, although a slight one. I didn't mind hanging onto it because that saved me the worry of catching one. There was a lot of coughing there in the barracks at Myles Standish. Taking a hike or marching from place to place turned the nose into an iceberg. I know if I'd ever smacked into a door I'd have been noseless, that's the way it felt.

"Twas the night before Christmas and all through the barracks,

Not a creature was stirring, they all had hysterics."

"Twas the night after Christmas and all through the barracks,

Not a creature was stirring..." they'd all been out on pass the night before.

Yes, Christmas came and Christmas went but to me it wasn't much more than an anticlimax to the warmth and cheer back in Springfield. The Service Clubs did a good job of spreading Yuletide Cheer, what with the shows they presented. Christmas dinner was good—turkey and dressing with all the trimmings and fruit, candy and nuts—and afterwards the officers distributed presents like books, cigarettes and chewing gum.

That evening we went to a dance at the Service Club.

Some lucky souls got to go home for Christmas because they lived within 3-day pass radius of camp—like Boston, Connecticut, New York or New Jersey. Lucky devils. Oh well, some get it, some don't.

To the rest of us so far from home, mail call took on special importance. After each one, the barracks was lined with GIs draped in unimaginable positions over their bunks, letter in hand, head in clouds. Then there were the ones who'd been missed at mail call, attempting to make up some feeble jokes, but their laughter dying away.

When the New Year came, I figured I really ought to get out and howl. Something like a rousing night club with serpentine and confetti flying all over the place, horns blowing, noisemakers making a deafening racket, yells and cheers, then the stroke of twelve, lights out, and a beautiful blonde kissing me... Not so.

You may not believe this, but I spent the New Year in the latrine. Another detail of latrine orderly trapped me in the last minutes of 1944, and there wasn't anyone around for company. So I wrote letters and made New Year's resolutions. One of them was not to get caught in a latrine next December 31st. And that was the way I ushered in 1945.

Since we already had an idea of how much spare time we might now and in the immediate future have on our hands, several of us decided to enroll in USAFI, the U.S. Armed Forces Institute, which gave us correspondence courses in almost any desired subject for a fee of $2.00. So I signed up for a course in elementary algebra, guessing that it might keep the kid out of mischief.

In the meantime, during the first week of January, we began to wonder when we were going to leave. We'd been here two weeks already, had received our equipment and had our physicals, our last shots, and all the routine, so now we must be Combat Serviceable enough to get out on the high seas. Still no word came down as to the sailing date, so we consigned ourselves to such amusement as was available at the time.

One such amusement was derived from an overnight pass on which several of us went to Providence, Rhode Island. We didn't get started till late in the evening in the first place and when we finally did get out of camp and hitch a ride, the car had a flat half-way to Providence and we had to stop and fix that. We got into the main Providence square shortly after ten and hunted around a while to see if there were any stray and gorgeous ladies who wanted to entertain GIs spending their last days in the USA. This late in the evening, we didn't have much luck. So we hopped a cab for one of the lower sections of town and got out on a dimly lit street. It seemed so odd that the city should appear to be in a dimout, but after walking around for a while, we came to a brightly lighted neon sign that advertised the Port Arthur Cabaret.

We climbed a flight of dingy stairs and went inside. A hat check girl took our surplus equipment and we proceeded on through the crowded room to find a table just a short distance from the dance floor. The room was well lit, there was an orchestra playing a low sultry tune on a beautifully lit platform across from us. Seemed like every table was full. Waiters moved in and out among the beams of golden light, serving soldiers, sailors, marines and civilians who had come to spend the midnight hours.

The menu came and we debated whether or not we wanted something to eat. We did. Being close to the coast, I thought it would be interesting to try some sea-plate delicacy to which I was not ordinarily accustomed. Shrimp? No. Clam chowder? Well… not this time. Halibut? No, I knew that pretty well. Something new. And there it was. Lobster. So we ordered lobster and champagne, just as if we could afford it. Just like kings.

While waiting for the lobster to be prepared, the boys thought they would have some fun. So they proposed the plan of ordering every drink in the house to see if there were any I might like. By this time, I had a wide reputation of not having ever become accustomed to drinking, and in fact disliking the bitter taste of beer, much less the rest of the bartender's lot. I simply had never imbibed much because I didn't like the taste, and that was that. But my associates found it very difficult to swallow that. Most of them could not imagine anyone so beyond the pale as not to take even a sip now and then.

When the champagne came, they reasoned, I would drink it down like a veteran. But in the meantime, wouldn't it be a laugh to get Shorty drunk for the first time in his life?

I was going overseas, maybe tomorrow. I would fight and die for my country and would never get back home again, so why not have a good time. Why not at last find out what this drinking business was all about? It sounded logical, reasonable. You only live once, so why not make the best of it? Why leave all the experiences of life undone?

So I said, bravely, all right boys, bring on the worst and I'll go along with you. I felt pretty well insulated against getting drunk. I would get sick long before that. The drinks started coming, and I started tasting. For most of them, only a sniff was enough. I tasted a few. Tom Collins wasn't too bad. Pink Lady, Martini, okay. I still remember the sauterne as being absolutely the most repugnant drink I had ever sipped. I had to wash out the taste with half a glass of water, and continued to retch for half an hour.

My friends gave up. They quit trying. My refusals had forced them to drink more than they should, and they were getting too plastered to enjoy the lobster. So we brought a halt to the drinking contest and prepared to receive the meal.

It came. A tempting, tantalizing lobster, claws and all. With it came the champagne. I dived right in. And then I dived right out again. By telling this, I risk being considered totally hopeless, a culinary boor, and a supreme idiot. I have no doubt that the lobster was delicious. The boys said so. But I had lived so long on dry land that my only meat was beefsteak. Or hamburger. It definitely was not sea food. I nibbled a little on the lobster and had to find some secret and unobtrusive place to spit it out. The champagne was even worse, and the bubbles didn't help. I sipped once and moved my glass to someone else's place.

I finally wound up having a French roll with butter and a glass of ginger ale. What a sissy!

The floor show began, and I forgot all about lobster, champagne and World War II. It was a fan dancer who wore the clothing customary of fan dancers. Every time the fan made a broad sweep, it would leave her momentarily in full view. My chair began inching toward her but the boys pulled me back. My ears went up and flapped a signal, my eyes popped out to the next table, and it was all I could do to keep from stamping on the floor or beating my head on the nearest table. I think my buddies were relieved that there was *something* here that I could like.

The dancer was, as they say, stacked like a brick penthouse, with all the bricks in the right places. She was a voluptuous blonde, and every time her fan made a wide sweep, a galaxy of GIs seemed ready to broad-jump the tables between them and her. Soon it was over, and I slumped down in my chair, out of breath. I looked around the room and saw that a hundred other soldiers had slumped down in their chairs, out of breath. Since we were too weak to absorb any more, we prowled around the streets a little more, then went back to camp. I crawled into my bunk at 0230 and had pink lobster dreams the rest of the morning.

On the 7th of January, bad weather set in again and it began to snow viciously. We were all called together and told that our date of departure would be the 8th of January so we rushed to the barracks to make everything in order and do the last packing. Passes were canceled, we were restricted to quarters. Night came and we sat around shooting the bull for a while, then hit the sack for a restless sleep. I lay there thinking of seas and ships and sealing wax and a lot of other things. Like watching the USA slink out of sight over a watery horizon...

And the night passed on.

NINETEEN
TROOP SHIP

The morning of January 8th didn't exactly dawn, it just got up to half-light and stayed that way. Snow the previous day had turned to rain and ice covered the ground. So the morning was bleak and dreary. But we weren't. It took us most of the day to get the short distance up to Boston. We rode down to the docks. I wanted to go up and visit a few friends on Beacon St. and the Hill, but my dinner jacket had been left at home.

We sandwiched the company on the loading dock, in a gigantic barn at the wharf. There were many other companies there and a lot of noise. We staggered in with our bags and parked in place to wait. Some stood around, others sat on the bags, some dozed. We could see the side of the ship through the big doors but couldn't tell much about it and we hadn't yet any idea of its name. Red Cross girls circulated around with boxes of candy and other knick-knacks to relieve the frustration of the moment. They were assisted by WACS, whom we ogled considerably as the last American girls we'd see in a long time. I asked one for a date and she said sure, honey, meet me tonight at the Coconut Grove. We adjusted our gear for more comfortable carrying. We were now individual units. Where we went, our household went. My barracks bag still had the two extra pair of shoes in it. I thought of dumping them overboard, but it would have been much easier just to let them stay where they were rather than to dig like a gopher to get them. Anyway, when we were torpedoed on the high seas, they'd make good lifeboats if I could get to them in time.

Eventually in late afternoon, the time came for all units of 1508-E to embark. We got instructions about going up the gangplank: our name would be called out by the officer at the end of the runway, and we were to repeat it as we stepped out of the USA and onto the boards.

At the signal, we hoisted our gear, stumbled a couple steps and lined up. The line went fast and eventually my name was called, I shouted it back and lunged up the gangway, fell back, lunged again, fell back….

That step off solid ground was rugged, especially since that solid ground was home, but soon I forgot about land and began to adapt myself to the sea. Step by step up to the deck and we could see that our ship was no mean troop carrier. It was a *luxury ocean liner!*

I could hardly believe it. The Matson Liner *Mariposa*, commissioned in 1932, now adapted for carrying troops. Well, this made us feel a little better. Just a lazy Atlantic cruise, that's all, and in real style.

As we were shown our quarters on the promenade deck, we suddenly changed our minds about style. In days of peace, this was the deck where gorgeous ladies in silk and satin and trimmed in lace strolled with nattily dressed handsome young gentlemen. They would lounge by the rail in the moonlight or sun themselves on a deck chair. Now it was not so. Nothing could look more in opposition to a promenade deck than what we entered. It was entirely filled with bunks, a double row directly down the center, four bunks high. There was a little space between our own personal bunk and the wall, but this was soon taken care of by placing all our possessions there so that there was no longer any hint of an aisle. Yet that was the way in and out of our quarters.

I selected a hammock third up from the floor, meaning that only one person would be sleeping above me, and threw my duffel bag in it to claim same as my property. It didn't occur to me until later that if the ropes of the bunk above me should suddenly break some night, I would be a mass of grease pulp. At the time, the bunk seemed ideal, for there would probably be a little more air up there and then, too, it would be at the head level of passersby, an advantage that would eliminate being kicked in the face like those in the lower bunks.

The place was dark and crowded. Oh, the lights were on, yes, but it took us a moment to find out where the lights were—

little dim things of about two candle power. Everything was closed in and the normal airiness and bright pleasant light of the promenade deck no longer existed. We were given instructions not to leave so we just hunkered around and talked. We got our blankets out and made the bed ready for the nights to come. Being so closed in, we thought, it could never get cold here, so we just got out a couple blankets. That was a mistake. Have you ever slept out on the promenade deck of an ocean liner crossing the Atlantic in the middle of January? I could tell you some things about that.

We were shortly given leave to go out on deck, and a number of us flocked out to wave goodbye to the good old Fatherland. When we got outside, it was dusk and a heavy overcast made it appear to be almost full night. From the ship's prow, we could see the Boston skyline, though few lights had yet come on. We looked below and saw the deckhands running to and fro. The last of the troops had long since been loaded and now the ship's crewmen were putting on the finishing touches and closing up the gangplank. This made us entirely afloat. It severed our last connection with land. This was the last chance to dive overboard and swim ashore.

The air was good out on deck, as contrasted to that below, which had become stale, so I stayed at the rail until long after dark—and watched our departure from a ringside seat.

Suddenly someone yelled, "We're moving!" I looked down. Sure enough, the dock slipped slowly by. It got farther and farther away. A blanket of darkness drifted down. Soon the wharf vanished in the mist of a January night.

Lights blinked on, in various colors, and moved into the distance, leaving a rippling trail behind. The water was calm and smooth. There was no wind. Soon, the sound of rushing water began as we picked up speed. We passed buoys, saw the dim lights and heard the tinkling bells. We heard a foghorn in the darkness, probably from a lighthouse on a distant headland. The lights of Boston blinked out on the far horizon and, shortly after, a small ship, apparently harbor patrol, passed us going toward the port. It began to flash a message, which we radio operators who knew Morse code, could pick up: "What ship?" We couldn't see the actual signal of our ship, but we looked up and saw a blinking glare of response. It read: "Troop transport." And then we were at sea.

The Empress Mariposa slid along the bay,
Slow, but getting faster—taking them away;
Boston lights behind—so carefree, so gay,
Came down to the water's edge to play.

They went upon the deck to think—
From out the darkness came a rapid blink:
What ship? Troop transport. Do not sink!
Then went below to catch a wink.

To say that the bunks had been built tightly together would be stating the matter with absolute truth. Neither the Army nor the Matson Line had anticipated that someone six feet five would occupy one of those suspended canvas cots. If we had all slept normally, which obviously I couldn't, our heads would have been about $1\frac{1}{4}$ inches from the feet of the man forward, and our feet would have been equally distant from the head of the poor guy aft. So we switched around and slept head to head, foot to foot.

The folly of that soon became apparent. I once gave a shove more violent than usual and cracked heads with my neighbor on the same level. I raised up in pain and cracked my cranium a second time, for not only were the hammocks crowded closely end to end, but if you raised your elbow a couple inches you nudged the back of the guy above.

It seemed hopeless. The only thing to do was to sleep in the same position all night long. Which was, of course, asking too much of my bones on that hard canvas. After squirming into the sack, I lay there a while and listened to the rush of water past the ship… and finally dropped off to sleep.

The *Mariposa* was at sea.

TWENTY
THE ATLANTIC

When morning came, we were awakened by loudspeaker. I opened my eyes and came to life quick enough to prevent myself from raising up. I lay there a moment and listened to the splash of water and spray outside. This was nothing like the quiet sound that had lulled us to sleep the night before. This was violent, angry, gnawing—as if the ship were submerging like a submarine every few minutes. I couldn't feel the ship sinking, though, so it mustn't be that bad.

But this roaring outside, these constant cataracts of water—what must they be? I rolled slowly over and looked outside, forgetting that the entire deck had been boarded up. Still, there were a few cracks, and by straining my half-open eyes enough, I could make out the ocean beyond. There were whitecaps pitching and tossing, but they didn't look sinister. I twisted over and looked on the other side, toward the wall. Nothing dangerous here, either. Then I looked at the floor and went white.

Little puddles of water glided across the floor. Rivulets of water ran in and out of our shoes, pants and duffel bags. I looked away quickly and stopped to collect my wits. This must have a normal explanation. Something was leaking, but where? Had it rained and blown in? No, I was dry. Had the roof leaked? No, the man above was still snoring and without gurgling. Had the deck been thoughtfully cleaned while we were asleep? (Now wasn't that a ridiculous thought?) After I had eventually probed all the possibilities and discarded them, I gave in to the last reason. The ship had to be sinking slowly.

I grunted and groaned to crawl out of the bunk, which was no easy task. At length, I flopped out and landed on the deck. I stood there a moment. Suddenly, the most awful sensation took hold of me—a paralysis that crept from toe to head. Every wheel in my head began to spin and all the gears started to clash.

Somewhere, some fool, it seemed, had begun to ring a bell. I started to see a few spots and then they grew in number, slowly turning from black to red. A feeling of somewhat faint familiarity began to spread throughout my stomach and I suddenly realized where I had felt that sensation before. On a roller coaster in the carnival. And so, quicker than you can say "Seasick!" I popped right back into bed. The wheels quit spinning and clashing, the bell stopped ringing, the spots disappeared and the funny feeling went away.

Now it was certainly going to be a helluva note to be pinned here throughout the voyage by something as silly as a funny feeling in the pit of the stomach. I looked around. A few others were beginning to crawl out of their sacks. So I mustered a vast amount of fortitude and plunged out of the bunk. This time, since I knew what it was, it wasn't so bad. As a matter of fact, it began to be interesting—like a ride on a merry-go-round. Quickly I got into my clothes, grabbed a towel, soap and comb and started aft. Here was my second mistake—thinking I could walk on a pitching deck.

Immediately I rammed into a post and staggered backward against a tier of bunks. I struggled to grab hold of my dignity and a post at the same time and slowly righted myself. Then I stumbled over a helmet that had been left square in the middle of the aisle and, as the saying goes, hit the deck... literally. I found my towel and soap and comb and got up again, this time proceeding with much more caution than before and after several minutes found myself in the latrine.

In front of the wash basin I had to devise an entirely new technique of performing the bathing process. I spraddled my legs approximately four feet apart and almost fell into the mirror. It would be necessary then to have a third point so as to become a tripod and withstand the assaults the floor beneath my feet was making upon me. So I grabbed the wash basin with my left hand and washed up with a single arm. I don't know that I got the slightest cleaner this way, and besides, the water was cold and salty, but it felt good anyway and I waked up a little more. Refreshed, I began to feel that my first day at sea was going to be okay after all.

Leaving the pitching latrine, I crept from post to post and sneaked along the wall, carefully planting one foot ahead of the other. At the same time, I had to make sure my head didn't bang against the low doors typical of ships like this. I thought I'd go out on deck for a moment and see what was going on. It was getting a little stuffy now that I'd been breathing for a while, and perhaps the fresh air would do some good. That old feeling in the pit of the stomach and that spinning in the head had worsened.

Outside I found a spectacle in progress. At least to a landlubber like me it was a spectacle. The entire world was in movement, with nothing anywhere in sight stationary. The wind tore the sea into convulsions, with turbulent, white-capped waves roaring up almost to eye-level then plunging back down into a foaming trough. The powerful wind turned the top of each mountain of water into a spraying geyser, whipping the white mists about in furious gusts. The sky was black, with low, ominous clouds. On the horizon, a thickening rain squall approached.

"Can't you just feel the great rhythm of nature out here, Dave?"

Spray hit me in the face as I bent over the rail to watch the flying water alongside the hull. The water and all the waves for as far as the eye could see were a dark, deep dirty gray-green, almost approaching black. As the ship lunged into it and turned up the water in a churning cauldron, I could make out a magnificent emerald green against the foaming white. There was a breathlessness about the whole scene that held me captivated, spellbound, witnessing a sight I had never before envisioned. My familiar landscapes had been mesas, plateaus, canyons, sharp peaks and thick forests, all stationary. This was a completely different experience.

As I watched it, I felt good, excited by the panorama. The last trace of seasickness disappeared and I felt it would never return. After breakfast I must return here and watch—for hours if necessary—and take a full look at the sea, its fascinating parade of changing waves and troughs in the roiled, gliding water. And to feel the thunderous pounding of the ship in the water, to study the frothing whitecaps that were being manufactured every second. Yes, I must get back here. But now it's almost time for chow.

I grabbed a rail and started back to the doorway. I was brusquely pushed aside by a GI coming pell-mell through the door, shirt half on, pants half buttoned, hair tousled and in a general semblance of disorderly conduct. He didn't care a damn what I thought of his looks. The only thing he could think of was getting the hell to the rail, which he did in no time flat, swaying from side to side. He bent almost double, and for a moment I thought he was going to plunge right on overboard. But he didn't go quite that far. His manner suggested that perhaps he might have liked to do so. His head went down out of sight, and I had a sneaking suspicion of what was going on.

I looked along the watery, tossing deck and saw several more rear-ends at the rail. While I was thus pausing, two more embattled GIs came plunging out to the rail, green-faced, and draped themselves over the edge of the ship as though to study the prancing waves.

Since I was feeling so good out here in the fresh air, I wondered if I dare go back inside, from which all this misery poured. I had begun to get hungry so I went carefully through the door, hoping to pick a moment when someone else didn't have the opposite idea in mind. Luckily, I made it in safely, and found myself in a dungeon where there was apparently no light whatsoever. I stopped and let my eyes accustom themselves to the change, and then picked my way to the bunk. There was the faint hint of a stench of sick men and I caught the dizziness on the way and had to lie on the bunk for a moment to collect my senses. I lay there until chow call came.

We had been issued two things aboard ship. One was a lifebelt that fitted snugly around the waist. It was in a collapsed condition and instructions were that twisting the little canister on the end would inflate it. We were required to have the belt on at all times except sleep and to practically go to bed with it. We were also issued cards permitting us to go into the mess hall for meals. The card was simply a tag with a large printed "B" on it, signifying we were from B deck. The loudspeaker opened up and a voice announced meal time:

"Attention, all troops on B Deck. All troops on B Deck will proceed at once to the Enlisted Man's mess hall by way of the midships staircase. Upon completion of their meal, they will return to quarters by way of the forward staircase. There will be no congestion of corridors or foyers during messing hours."

These, to me, came to be welcome words.

In order to eliminate continuous confusion on board, meals were served only twice a day… breakfast and supper. That first morning I made my way haltingly down the midships staircase and, guided by sentries posted along the way, came into the mess hall and joined the line. We picked up our big metal trays and eating tools and began to receive the meal, cafeteria style. I looked down the line at what they were offering and began to wonder if I would make it: not just the ordinary bacon and eggs and prunes and toast, but beans and wieners, thousands of them, in a great big vat from which our portions were being meted out.

I managed to look, though, while they filled my plate with generous helpings. I figured the large servings were good because this meal had to last all day. With loaded tray, I stumbled across a pitching floor and barely made it to a table. I sat down and suddenly realized I was indeed hungry.

While eating, I looked around. This may have been a mess hall at this particular time, but in the good old days it had been

"Don't ask."

a dining salon for the elite aboard ship. Long rows of temporary tables had been installed to accommodate a number of men. The lower walls of the room had been boarded up and I assumed they protected mirrors or paintings or some such delicate structure. Above, indirect lighting illuminated the upper walls and ceiling, where exquisite paintings depicted seas and ships and magnificent seascapes. All that improved the atmosphere, psychologically at least, and in a short time I pushed back from a solid meal, feeling much improved.

I returned by the forward staircase and went to my bunk to put things in order so that I could go roaming about the ship for the rest of the day. It was bad on the promenade deck. Many had already been sick in their bunks, too sick to move and unload on deck. So they had committed their sins in the aisle and others were busy at work cleaning. I heard groans of pain and misery and voices pleading for a quick death to end it all. Vic Roberts, in a bottom bunk below and aft of me, appeared to be having it as rough as anybody, cussing like a veteran and holding his sides.

"Aw, come on, Vic," I said. "Get up and go down to chow and you'll feel better. They've got good old scrambled eggs and with crisp bacon, delicious pork and beans, and fresh boiled wienies..."

Vic writhed and broke in. "Shut up, you bastard! Go away and let me die in peace!"

I went away, and, being a true landlubber who hadn't even *seen* the sea until arriving at MacArthur, I found myself drawn back upstairs to the rail, watching the waves, the clouds, the sky... there was more than my two eyes could take in all at once. Now and then, porpoises leaped in the sunlight. I studied the waves in great detail, tossing and rolling, a million whitecaps on windy days, lazy swells in calmer weather, the foam, the light aquamarine colors, the depth, the darkness... I compared the troughs and crests of the waves to gullies and hills in the Painted Desert, the swishing spray to the waterfalls on Oak Creek in the red rock country around Sedona, where I had camped many times.

I came up at night, after chow, and on clear nights thought the celestial displays some of the most beautiful I'd ever seen. Venus held a prominent place in the evening sky, more brilliant than I'd ever known it. I thought of it almost as a second moon, because its radiance reflected across the waves. One night I stood there with a friend, who said, mirroring my own thoughts: "I could just stand here for years looking at that, and relive every moment of my life."

The first three days of the voyage were spent in storm, a rolling, pitching introduction to ship travel for many of the boys on board. So they stayed inside. Some remained in their bunks the whole voyage, except perhaps the last day or two. But after the storm, a calm sea carried us the rest of the way. The weather warmed up, too, even to the extent of allowing us to play cards out on deck in the open air. In time, fresh air and scrubbing gradually cleared away the foul stench on the promenade deck, where we slept, and most of the victims resident there got over their seasickness. Still, being in the hold, closed in, surrounded by darkness, induced dizziness, so we spent as much time as possible on the outside.

I frequently rambled along the deck, and in the passage-ways where enlisted men dared to tread. The bow of the ship, to me, became a delightful place to stand, to surge forward and rise with the ship. All dizziness vanished as I became tuned in to the motions of sea and planet. I looked out across the water, felt the sweep of fresh salt wind in the face, and looked up at the bridge, where the crew operated a floating village.

Along the walks at the side of the ship, port or starboard, you were closer to the water and could observe the waves and foam beating against the side of the vessel. Astern you bounced in rhythm with the propellers and watched the foaming wake of the ship stretching far behind. Most of the time, the ship's trail was zigzag in the water, for we were going across unescorted and this process we assumed would assist in avoiding enemy submarines. Being so unescorted gave us some concern and we resigned ourselves to being torpedoed and burned at sea, but so what? You gotta get it some way. It didn't help to hear of the sinking of the transport *Leopoldville*, a Belgian ship being used by the Allies. However, other news bulletins we received from the ship's press heartened us: the success of the Russian offensive and the Philippine campaign.

Then, trouble loomed. Ocean travel is known to interfere with the normal physiological functions of the human body, and I proved it. I was constipated the first four days. I had no urge, no capability, no success. It was as though I had dried up inside. That was definitely false because of the giant helpings of food I gulped down. I began to wonder where all that largesse had gone, then had to confess that it hadn't gone anywhere. It was still there, just waiting... waiting... growing larger like a boulder in my midriff.

At the end of this time, and in desperation, I put to myself three choices: (1) Drop dead; (2) Explode; or (3) Sit on the john and *stay* there until *something* happened. Adhering to the law of self-preservation, I chose the third option. The whole thing was beginning to affect my enjoyment of the trip, consumption of food, and ability to sleep. I wasn't sure I ought to eat anything more until *something* happened. So I determined to settle the matter once and for all, and retired to the latrine aft.

Of all the places on the ship to select, I had to pick the one that sat directly over the propeller. The whole room bobbed up and down in a vibrating rhythm. If you have never experienced this, you may not be able to imagine how such a motion inhibits bowel movement. One is supposed to relax, not grip the seat to avoid falling off. It is even difficult to piss when you have to brace yourself against every shudder of the ship.

I was not going to give this up, however, no matter what. Nothing could interfere with my determination. I had come here to succeed and I was going to do it all by myself. No enemas. No mineral oil. No Nature's Remedy tablets. If I wasn't big enough, resolute enough, to solve *this*, how could I solve any serious problems in combat? That was something of a specious argument, because few problems in combat compare to the inability to shit.

So I sat and meditated on the seas and the stars. I squirmed and shifted on my seat from port to starboard, fore to aft. I concentrated, striving mightily to exclude every vestige of thought from my fevered brain. I hunched over. I sat up straight. I stood and then sat again. I verged on delirium. Where *had* all those beans and wieners gone? Would I have to have a Caesarian? Dynamite? This may seem frivolous, but I can assure you that I was exploring every avenue of rescue before keeling over and falling onto the latrine floor.

At last, I began to vibrate with the ship, and finally, with the utmost diligence and effort, I succeeded in my mission. I had never known such relief. From then on, I returned to normal. At least in combat, the slit trenches wouldn't be vibrating. It was suddenly a brave new world.

We played cards, day in and day out. Most of the time we squatted around the compass on one of the rear decks. It was crowded with other games going on, but we usually found room to spread out a blanket and so pass the time. Pinochle was the commonest species but we switched to bridge from time to time just for the change. We also spent time in studying the little French and German language books that were distributed shortly after sailing. Occasionally, PX items came through and were lowered down through the echelons until arriving in our section, where we purchased candy and gum and cookies and other snacks. There was the ship's library, too, and we checked out books for casual reading. We had lots of time. I read a great deal, and relished it immensely.

One morning the scuppers got stopped up. These, we were given to understand, were the vents through which water drained from the decks. Water had somehow gotten into the deck and we found about three inches of it in the aisle and under the bunks. Some of the boys had their canvas sagging so low in the lower bunks that they waked lying in water. The mess made a horrible sound, too, when the water swished from one side of the deck to the other, accompanying the roll of the ship. The victims so wetted growled a couple of times and hauled themselves out of their bunks. Well, you can say one thing about our trip overseas on the Mariposa… there was never a lack of humidity.

One of the most haunting questions that plagued us on the way was where we were going to land. Africa, England, France, Norway, or right on through the Mediterranean and in to the Indian Sea—we had a lot of guesses. One of them, though, was right. The answer was Marseilles. So we got out our little French books and with thoughts of little French sugar plums dancing in our heads, we began to take a new interest in the language.

Marseilles! The gem of the Mediterranean! The warm, sunny, inviting Riviera—its resorts, its sun-drenched beaches and all the thrills and pleasures of life abroad. I could just see it now—wiggling my toes in the sand beside the blue, sparkling Mediterranean. Perhaps a junket to Oran, or Algiers, an exciting camel ride across the gleaming, burning sands of the wide Sahara. And certainly there would be time for a trip to Naples, to Rome, to all the fascinatingly beautiful spots that swam in history. A side trip to the romance of Capri, by sailboat past the sinister Volturno, or a weekend of yodeling in the Alps. There was *everything* here to do. I didn't do one of them, but how I could dream!

At least, landing in Marseilles would permit of one thrilling sight along the way. This was the Rock of Gibraltar. And we went by this at one o'clock in the morning on a cloudy night.

A day before landing in Marseilles, we waked to find a destroyer escorting us. This was the first ship we'd seen on the trip, with the exception of passing the tremendous luxury liner *Ile de France* our second day out. Then night fell and it was announced that we'd be going through the Straits at 1:05 a.m. Bitterly disappointed, I got up anyway and went out on deck. At least I can say I saw the Rock. One section of that pitch blackness out there was it, but I sure couldn't see a thing. Except, that is, for one red light and one green light out on the water.

Even by the time we passed from the Atlantic over into the Mediterranean, there were many men who had not been out on deck. Poor Vic Roberts was one, or at least it was a long time before he saw the light of day. We had a fire drill one day, after being carefully instructed on how to file out and exactly where to go. When the alarm sounded, we all began scooting through the aisle. I stopped momentarily and asked him if he were coming along, and I got a curt answer.

"Go to hell!" he said. "Let the damned ship burn."

TWENTY ONE
DELTA BASIN

On the 17th of January, we arose to cries of "Land! Land!" and went out on the open promenade deck on the other side of the ship to have a look. Sure enough, there to the south, under the gleaming sun and across the sparkling waters lay a low coastline stretching for miles in both directions. Africa! We watched it longingly, not so much that it was Africa, but that it was land… good old terra firma right there before our eyes. The romance and adventure of the ocean voyage had by now worn off on our 9th day at sea, and we welcomed something new.

GIs crowded the deck to the hilt, and I'm sure the ship must have listed several degrees with all the weight having been transferred to the right side. But the view was spectacular and you couldn't blame the boys for wanting to drink it in. I was standing near a fellow with binoculars and he let me use them to get a closer look. We passed two North African cities which we calculated to be Algiers and Oran. They were sitting down close to the water, with a rise of land behind. We could see the buildings, the ships in the harbor, the smoke of activity rising in the early morning sun. It was habitation, human habitation, and it looked good. We were getting tired of looking at sea and ship and soldiers all day long, so we took this in, all and entirely. About this time, the horrible thought struck me that if we'd gone over the Pacific way, as I had wanted, we'd have been at sea at least three weeks instead of one and a half. Gad! Perish the thought…

I looked at the blue Mediterranean and wondered where they'd taken the blue. It seemed no different than the black Atlantic, and half-stormy weather chopped it up a little. Maybe some of the shores on the Riviera were warm and sunny and the sea was blue but certainly not here. I gave up trying to make anything out of travel advertisements and went inside. We were told that the ship would dock the following day in Marseilles, so we began to get our gear together and make ready for landing. I thought I'd break down and take a shower, even though there was only the promise of salt water—and cold salt water at that. We had been issued special bars of soap that would supposedly soften the water, but I scrubbed for half an hour and never did seem to work up the right kind of lather. I got cleaner, though, and even shaved, so I felt more ready to land on Europe for the first time.

A couple of us wandered around the ship watching other units preparing for the day to come. We went down on C, D, and E decks, below water line. We went up on A Deck and visited with the boys whose bunks had been built on the Mariposa's dance pavilion, and it wasn't bad. We came to the conclusion, though, that our promenade deck wasn't bad either. At least, the air was fresher.

Next morning, the only thing in sight was the blue-green-black Mediterranean. It was cold as we stood on deck waiting for land to show and there was an icy breeze, whipping white-caps off the waves. Almost everyone was on deck, it seemed, for this was indeed the big day—so very few had ever been to Europe before, fewer even to have seen Marseilles and the Mediterranean coast. Now we were preparing our positions from which we might gain a view of the land that was to become our home for a considerable amount of time to come.

And then, before the sun had moved a couple inches in the sky, a low, rugged, unmoving line appeared on the north horizon. We were going northward slower now and began to see debris floating in the water. The destroyer that had picked us up the day before Gibraltar had only escorted us to the straits and now we chugged into Marseilles all by our lonesome. As the land grew nearer and we began to pass headlands, all necks on board ship extended like giraffes,' all eyes popping to see, and countless questions as to what is that headland, that bay, that village over there—and just as many I-don't- knows. Coming into Marseilles was a beautiful experience, that is to say, to us it was a morning filled with color.

We passed some kind of stone marker on a point jutting out to sea. There were some small ships that had come out from the harbor, fishing perhaps. There were stone shacks and huge company warehouses on barren flats that came out from land. When we got into the harbor, we scarcely inched along in the water. There was little but destruction in the harbor—the masts of sunken ships, each bearing the undying Tricolor of France.

The sea took on a bright blue that matched the sky. In between, there were the contrasting colors of the city, the brown earth and brown-cream cliffs that broke away behind and rose above Marseilles. There was a highway passing along the coast to the westward and rather than pass along the contours of the talus or natural slopes of the cliffs that came down to the sea, it bypassed the curves and straightened out with a series of long, high graceful bridges with long arches beneath. The city itself bore the bright green of trees (though how anything could be green on a cold day like this I couldn't figure out) and the shining red of tile roof tops. It was a big place, whose levels stretched alongside and above the harbor all the way around. This then, was Marseilles, city of Mediterranean intrigue and international crime, city of culture and of France—now of damage, waste, and ruin, yet hustling and bustling. Landing port for American ships—like the Matson Liner *Mariposa.*

It took most of the day of the 18th to unload the troops. "All personnel of unit 1508-E will prepare to disembark." Several of us from 1508-E did not disembark that day, however, for we were placed on the packing and unloading detail that would take charge of assisting in taking off some of our equipment. From our vantage point at a port hole we watched the rest of the boys hop down off the gangplank into France. They lined up along the wharf. Trucks kept coming to load them in and haul them away through the streets of Marseilles. We wondered where they were going.

In the afternoon we didn't have much to do and were told to get a little rest before the work of the evening. A few of us toured the deserted ship. We went out on deck and took in the closer view of the city. The streets we could see were, of course, cobblestone. Not a wooden building could be seen. Uptown rose the larger buildings. Along the harbor we saw gigantic submarine pens, now destroyed, and we could see *le vieux port et le pont transbordeur*, though only the towers of the transshipping bridge remained intact.

In early evening, we were called to the wharf and began assisting with the unloading process. Here we got our first glimpse of German prisoners of war, who were working steadily, though somewhat sullenly, for to them, the war was far from lost. We worked on into the night and continued by artificial light. I noticed the narrow plank over which the boxes and crates were being hauled and I noticed the small area of open water between wharf and ship. Sometimes it was a space five or six feet wide, at other times the ship nudged gently right up to the wharf and I thought that if anyone ever fell in he'd better select a time when the ship moved outward or he'd be crushed to grease.

While I was thus thinking, one of the men slipped and fell in. This was disagreeable enough, cold as the night was, and cold as the water must have been. I suspect it was a lucky thing he knew how to keep himself afloat. Nevertheless, we moved like lightning and lowered one fellow down to the water by his legs so that he might grasp the swimmer below and pull him up. If the ponderous hull of the ship should make a movement toward the dock before the rescue could be completed, both men would be lost.

Actually, it took only a few seconds but the time seemed interminable. Eventually we pulled and strained and got them both safely out on shore again and color returned to our faces. The poor guy who had fallen in stood shivering, probably from both cold and a close brush with the reaper. Oddly enough, and this we could laugh about with him, he still had his eyeglasses on. He was bundled up and hustled inside. I tried to dive in so that I could be hustled inside where it was warm, but they stopped me. So I finished on the detail. I'm kidding, of course, but I thought about it.

Our work ended at two a.m. and we went on deck to snatch a few winks. Next morning a truck came and got us Signal boys and our gear. We asked the driver where the company was stationed, right here in Marseilles maybe? No, he said, it was a place called Delta Basin Staging Area, or DBS for short, and was something like fifteen miles from here. We loaded up and took off through the streets of Marseilles.

Almost immediately we got our first introduction to the *laissez-faire* of French culture. We had been told it happened in France, but we didn't believe it. Right there in broad daylight, in the morning sun, stood a middle-aged man in gray cap and black, disheveled suit. He had been briskly walking down the street. Suddenly he stopped, turned to the gray wall beside him, unbuttoned his trousers, and relieved himself. I couldn't help thinking of an American dog coming to a fire hydrant. There weren't many other people on the street, and if the street had been as crowded as Fifth Avenue, no one would have noticed. The fellow did the whole thing so matter-of-factly. It astounded us, and as we drove on we noticed him finish and saunter down the street, as if nothing had happened.

You may think our reactions a bit odd, but it was the first time we'd actually seen a practice so common on the Continent. We had been raised in a long line of citizens who confined their pissing to a toilet, if for no more reason than public hygiene. We realized that in Rome we must do as the Romans do. But I felt that if that applied to France, if that was the only place to relieve oneself, I was destined to die from extended bladder. I can appreciate quaint customs, but for the American GI this was a bit too quaint.

The streets of Marseilles hummed with activity. Stores seemed to be doing a brisk business. People, many clad in gaily colored coats, walked the sidewalks. Others rode bicycles. Still others pushed or pulled carts through the streets, the wheels clacking as they bounced over worn bricks. Some American GIs were not known for their couthness. Whenever they spotted "gorgeous babes," as they called them, they howled, threw kisses, and yelled: "Bonjoor, madda-mo-zelle!"

They often got a response. (Liberating armies are forgiven much.) "Ah, bonjour, soljare!"

"Oui, oui, bébé." Smack. "Voulez-vous couchez avec moi?"

From a moving truck, of course, the chances of sleeping with the girls one passed by was virtually nil. The girls would sometimes turn away red-faced. Asking the resident beauty queens to sleep with one was not exactly a diplomatic statement, but it so characterized the GIs that came to France that it became well nigh universal. Even the sorriest on language capability learned a few key phrases, fast. To give a little credit to the GI, the phrase was often invited, if not stimulated, by coy gestures, prancing walks, and eyebrow movements on the part of the demoiselles addressed. Little wonder that the troops responded. The GIs viewed this as a form of free entertainment, available right there on the street, a means of thanking the liberating soldiers…

We wondered if we would ever get into the city on pass and tour around and see the sights. We were told that a number of places in the lower section of the city were off limits to Allied personnel. As a matter fact, whole sections of the city had been roped off for Americans. But that didn't bother some of the men, if we were to believe the stories of fantastically exotic evenings spent at a certain address on Rue Madeleine. Rue Madeleine was, of course, off limits, but that made it more romantic and the scuttlebutt was that you could find better girls there. I hasten to make clear that I am not speaking from personal experience. I never got in on a sight-seeing trip to Marseilles. However much I may have wanted to, I could still remember the vivid training films that depicted closeups of social diseases. With these fresh in mind, I limited myself to enjoying the tales brought back by more daring and adventurous members of the team.

We rode up along the cliffs that overlooked the port city and finally into the valleys beyond. At length, we arrived at Delta Basin and rode what seemed like miles along the encampment where pup tents were stretched out on both sides of the road. We made out a number of temporary buildings and a few Quonset huts that we assumed to be company headquarters, service clubs, and so on. It was a broad valley, almost a plain, rimmed by low hills. Roads everywhere, in checkerboard pattern, were soft and rutted. Finally, we rode into the 570th Signal Company area.

Now from here on, the directions I give you will be entirely fictitious, for I lost all sense of compass direction. Sure, I could tell which way Marseilles was, which way to the mess hall and latrine, but as to which was north and south, there was no proving. I couldn't tell by the sun, since it rose in the wrong place and half the time didn't rise at all. An overcast characterized our visit to southern France. By this time, I began to lie in ambush for the so-and-so who told me about the beautiful, languid, sunny Riviera. There was cold, and ice and wind—and all damp, just like the previous winter in Alabama, so that it knifed into you and kept you cold all day long and half the night. But we were at war, so take it, soldier, and quit griping.

The mess shack sat at the north end of the company area, along with another building that housed the administrative officers, buildings as temporary as you might imagine. Chow was served from a bench outside and now our mess kits and we came to be very good friends because we saw nothing of plates for a long time to come. Cans for washing our mess gear were placed near the serving area and SOP (Standard Operating Procedure) called for three of them, one soapy and two rinsing. They were large garbage-type cans with little oil stoves that dipped down into the water and heated it.

The latrine lay across a country road at the south end of the area. This structure was nothing more than some logs thrown up and covered with canvas on three sides. As for a place to wash and primp, well, that was taken care of by our steel helmets, which, for weeks and months to come, would perform the functions of boudoir, vanity, and bathtub.

We got no baths while in Delta Basin. Our tents were pitched between the extremes of mess shack and latrine, with a "company street" down the center. This was our work and training ground, and a place where we built fires during the day and evening to get warm—extinguishing the fires before nightfall for security reasons.

Dody Ratigan and I pitched our tent, slapped in our duffel, made beds and called it home. Outside, we dug our foxholes and with that established residence. All the time, the weather was miserably wet with rain and cold with ice.

We spent the first couple of days getting settled. There were a few work details until everything was secured. We turned in our American cash for French money; I traded $8.16 for 404 francs. We got used to our new food. We had some K and some C rations which somehow had improved over the types we'd received in basic training, or maybe it was just the locale that made them taste better. Regular chow was good and plentiful. Coffee was served for two reasons: to drink and to warm the hands with. I didn't drink coffee, but a canteen cup of it felt very welcome to my freezing hands each morning. There was some dysentery until the cooks got used to combat conditions, but so little as to be a tribute to Norman and his staff. He did a good job for us.

We found they weren't kidding when they said: "Wherever you go… the American Red Cross will be there…" It was, even in the DBS area. "You are going overseas," they said, "on a—well, let's call it, grimly, a business trip. We can't tell you anything about that. But we can tell you about a few things going on overseas that you may not already know. We mean the things that will lighten a little the hard, slogging work you have ahead of you. We mean the services offered by the American Red Cross.

"Take clubs. Like a snack with good coffee? Like dancing to sweet music? Like lounging with your favorite magazine? Like showering off that gray dust—and then sleeping between sheets? When your time is free, look up your nearest Red Cross club. Hundreds of clubs and rest centers dot a cheerful line clear around the world…"

We had a club and it was cheerful, but it didn't have dances, showers or sheets. Nevertheless, it was attractive enough to draw us over two miles by foot almost every evening. They had movies, games, easy chairs, phonograph records and lots of writing desks. There was coffee and cookies and cake next door for 5 francs (10¢). If we didn't want any of these services, we just went down so's we could remind ourselves of what it was like to sit in a chair once again. Most of the daytime we squatted, and that's all right if you like squatting.

The weather continued so cold that the water in our canteens froze overnight and didn't thaw out until mid-afternoon the following day. I waked one morning and it looked like Canada outside, with a light covering of icy snow on the ground. The first night or two, we got cold sleeping, but then we wised up and made our blankets into a kind of sleeping bag and were warmer from then on.

One of the big troubles with setting up household for two in a pup tent was that it left little or no room for us. We crawled in and left our shoes outside. We squirmed and twisted and squeezed to bury ourselves in the covers. We often thought of what a predicament we'd be in if there ever came an air raid. We did have an alert, but nothing came of it and fortunately it was in the afternoon. We spent a few minutes in our foxholes and

then crawled out sheepishly when someone said it was an American reconnaissance plane. We had a red alert one night, however, but we didn't know a thing about it, dead to the world as we were. It was a German recon plane photographing Marseilles.

Being so cramped in the pup tents brought about a serious situation during rain. The shelter halves were of that type of waterproof canvas which was waterproof until touched on the inside. When we crawled in under these conditions, it was all that much worse trying to squeeze around without touching. Dody and I learned to take turns crawling in. When I went in first and struggled to get in bed, Dody would stand outside and say, through chattering teeth: "Hurry up, damn it! I'm getting soaked out here." Hurrying up was the last thing anyone could do, so he got soaked. When it was his turn to crawl in, I did the same thing and I, too, got soaked.

A delicate situation developed in Delta Basin that the GIs ironed out in their usual way. The piercing cold of the days and nights had a tendency to require nightly excursions down to the latrine. Some men would have to get up several times during the night, but they didn't do this long. After a time, you could see steel helmets reposing just outside the pup tents all the way down the line. These makeshift urinals were used grumblingly during the wee hours of the night, so to speak, emptied in the morning, washed out a little, and worn during the day. We always got a laugh out of sleepy-eyed troops parading down to the latrine each morning, helmet in arms, trying to run because they were in a hurry, yet daring not to run for fear of spilling the contents of containers they were carrying.

I was one of the luckier ones with sufficient capacity to sleep all night without being interrupted by nature, so I felt secure to wash in my helmet and wear it without thinking of the night before.

Not so Dody. On one of the coldest nights, he filled his helmet and set it outside the tent. Next morning, he found it frozen solid. Try as he might—whacking, banging, and thumping the helmet on the ground—he couldn't dislodge the contents. As the time for morning muster approached, he got worried. To appear at muster without a helmet was a serious offense. Not to appear at muster was an even more serious of-

"I couldn't get the ice out."

fense. We looked down the line and saw other men desperately banging their helmets.

Dody luckily freed his helmet just in time, setting the ice by the tent for disposal after muster. No chance of it melting until then.

From this, I drew Enduring Lesson Number Four: *If you can't hold your piss, don't go to war.*

"Hopkins, did you go into Marseilles last night?"

TWENTY TWO
PROVENCE

The first six days at Delta Basin were comparatively lazy ones. The big factor apparently holding us up was trucks. We had the radio equipment, but nothing in which to install it. So we lounged around, waiting for the trucks to be delivered.

This gave us an excellent chance to get acquainted with the surrounding countryside and to visit some of the people. Fred Larsen was the unofficial interpreter for the company since he had taken some four years of French in a New York college and spoke it fluently. I had my high school Spanish and wanted to try it out, figuring that this being down near the Spanish border, there would be some who knew both languages. So one evening, Fred and Dody and I set out to see France.

We wanted to get a good close look at the countryside, and found it to be a series of villages liberally sprinkled with farmlands and vineyards, all connected by picturesque, tree-lined roads, much resembling our own country lanes. Another thing we wanted to find was French bread, with perhaps some wine. And this is how we found that the war had smashed France.

After hiking five kilometers (close to three miles) we came upon a deserted-looking group of farm buildings. The desolation, the tumbled-down haunted appearance we at once assumed to have been the end result of Nazi pillaging, and we found we were right. In peace time, the house and yards must have been kept up beautifully, with exquisite landscaping and gardens... and even with magnificent architecture. The decay of the occupation years became obvious everywhere. We saw no domestic animals except a few goats and perhaps now and then a dog or a cat.

We knocked at the door and soon found ourselves looking into an extremely pathetic face. It was a woman in middle age, unkempt, dress dirty and hair straggling over her ears, yet of an odd beauty, a simple beauty.

"Avez-vous du pain?" I asked.

She didn't look at me but one or two quick glances. She stared mostly at her feet. She seemed listless, all the spirit gone from her. She spoke in a weak voice: "Non, monsieur."

I thanked her, gave her some cigarettes for her trouble, and stepped away. That was that. I felt a tinge of pity, not only for the suffering that was apparent in her face and in the empty, darkened, perhaps unheated, room I could see behind her, but for being unlucky. I could envision walking up to a farmhouse in the USA and being greeted by a woman who might turn up her nose at the thought of giving or selling bread, who would have in her house a bountiful larder. Yet she might never in her life have appreciated how lucky she was to live in a free land that had never felt the scourge of a total war and occupation such as this. I wished some of the people in that far-off country could have read from this French woman's face what I read from it... The very sight might have made more palatable the burdensome wartime taxes, and the rationing of familiar delicacies that could only be enjoyed once a week. How I wanted to help, somehow. I *was* helping, just by being here. But I couldn't change the world overnight. So we picked up our heels and found better foraging down the road—bread, jam, wine, for which we paid handsomely.

At one point, the lady who opened the door stood looking straight into my belt buckle. Then she slowly lifted her gaze, looked up into my face, and said: "Que alto! Que alto!" She spoke Spanish! "How tall I was!" And what a wonderful sense of humor she had. I was overjoyed to be able to speak with her so easily in Spanish. Her name was Madame Joan. She invited us in to munch on the bread and jam we sought. She felt shy to tell us much about the hard times they'd had, and I felt that she had tried to drive it from her mind. She wanted to be positive, asking about our homes and families instead. It was an hour before we could get away from there, lavishing her with gifts as we left.

The cold continued and canteens kept right on freezing. Everyone thought I was crazy in my plans to include a lot of camping when I returned to civilian life. Until I told them I'd do my camping at the Equator.

On the 25th of January, a group of us got off on pass and, since we could not go into Marseilles, decided to roam the countryside again, and this time we did a more ambitious job of it—a round trip of about 15 miles. We hiked to the little village of Les Milles, not far from the University town of Aix-en-Provence. We enjoyed the hike in the cool of a cloudy afternoon. We passed farmhouses all along the way, crossed and recrossed railway tracks and passed an abandoned airstrip. We stopped a while to talk with a family in front of a house with a brick porch.

In Les Milles we found a little café where we ordered bread and wine and sat in a booth to consume it. I asked if there were something other than vin rouge naturelle to drink, and if anyone around here had ever heard of Coca-Cola. The woman behind the bar looked at me as if I were some subversive agent. But she brought out a big bottle of *limonade*, a lemon drink that tasted like 7-UP.

We watched the cafe family having dinner. It wasn't a swank affair for the only staples appeared to be soup and potatoes, but at each plate, children and all, sat the inevitable bottle of wine. I don't think I ever saw in all of France a drinking glass filled with water.

We sat and talked and laughed as dusk crept in on Les Milles. I killed my first bottle of *limonade* and called for a second. When I'd finished that one, I felt I had better go outside, so I went around into the darkness of the alley for a minute, then came back and felt able to order yet another bottle of *limonade.* I think each bottle held about a quart. The other boys got pleasantly happy on vin rouge naturelle.

"Yes general, we find the French very cooperative."

The village was not big at all, and there were only a few stores open at this time of night, but we found a certain fascination in walking along the quaint streets and observing the few people out at that hour. We dropped into a little shop where my colleagues purchased a bottle of cognac and I fascinated the attractive young waitress there with my red hair. We got along well because she had red hair, too. I could have stayed there several days, but the boys insisted on going back to camp and I didn't want them to go alone. So I bade a fond *adieu* to the ma'mselle with the red hair, and we took our leave.

The long hike back turned into an ordeal. Anyone who had gone on maneuvers way back in Morrisville, Alabama, could have survived this sojourn in Aix-en-Provence. But it was still a long, long hike at midnight. By one-thirty a.m. we had our weary bones back at camp and in bed.

Thus we got our introduction to the sights and the scenery and the people of southern France—and our prelude to the ETO (European Theater of Operations). I found it fascinating. It would undoubtedly get more so as time went on.

And more dangerous.

TWENTY THREE
HEADING NORTH

Not much of January remained when the trucks arrived and we began the job of mounting our radio sets in them. The SCR 193s had to be installed, with accessory equipment, in weapons carriers $^{3}/_{4}$-ton, and the big 399 had to be prepared for the long haul up north to the front. We were given to understand that time was short and we must get all the vehicles and their equipment combat-proofed in the quickest possible time.

It involved working hours that led on into the darkness of night. All sections of the company had their own work to do and everyone was busy from dawn to dusk. Our own particular work involved drilling mounting holes and placing the radio sets up against the front of the truck bed, then tightening them down as solidly as could be done so that they would stand the rigors of combat through which they must soon go. When the transmitter, receiver, loudspeaker and tuning unit were installed, the converter and auxiliary battery were placed and the wiring began, both to supply the sets with power and to shield the motor so that it would not interfere with communications during mobile operation. As much as possible, the truck was weatherproofed, for we still had some winter to go. Canvas tops were tightened and sealed down.

I had to wait until this most inopportune time to catch influenza. We had received flu shots back in the States, and I felt they had done a superb job of keeping me out of a severe cold, but apparently this sleeping in continual cold and on the damp ground would eventually take its hold, so I came down. However, I wouldn't admit it. This was no time to become bedridden with so much to do and so short a time in which to do it. So I kept plugging ahead with the rest of them until one afternoon I was drilling some holes in the side of the truck and suddenly turned so weak I couldn't press the drill against the metal. I kept trying, but it didn't seem to work and I asked Luke to give me a hand. He came over and started to help, and then got a look at my face.

"What's the matter with you?" he asked. "Your face is yellow as a lemon!"

I told him to forget it and start drilling the holes, that I would guide the bit into place. But like a fool, he called some of the other boys over and then one of the sergeants gave me tender but firm orders to get the hell up to the dispensary and see what was going on. I grumbled once or twice about the work left to do and this being no time to slack off, but they kicked me out of there and I went up to the dispensary.

This was a field station located in an Army tent. There the medics did little more than swab out my throat, give me some kind of pills and tell me to get to bed as soon as I got back. So I staggered back to the tent, pausing first to cast a longing eye down toward the end of the company area where everybody was busy in the excitement of getting ready for the big push. Then I crawled in, muttering a thing or two under my breath, and buried myself in my sleeping bag. I stayed there all afternoon and all night without getting out.

Dody brought something as a supper snack, but I didn't care for much food. I kept the covers tight around my neck and some over my head so that I would remain hot and perhaps sweat it out.

This happened to be exactly the proper method, for the following morning I crawled out of bed well again. Weak, perhaps, but up and at 'em and ready to go for the last day of preparation.

We were constantly reminded of the prevalence and danger of trenchfoot, a horrible disease that resulted from getting the feet damp and walking around all day that way. Of course, we never allowed our feet to air out much because of the bitter cold, but we did manage—in fact, were ordered to—keep fresh, dry socks on at all times. There was mud and this required use of shoe pacs, rubber boots that allowed no air circulation whatsoever to the feet. Thus a great deal of perspiration took place about the feet and kept them damp. From the front lines came stories of doughboys who had gotten trenchfoot and it had taken them out of the war. It was not an especially pleasant method of being taken out of the war, though, for severe cases required the amputation of the foot or leg.

With this little thought in mind, we dried our feet off night and morning and wore dry socks during the day. We learned the symptoms and kept watching for them so that the disease could never get a start. For a long time at Delta Basin I had one of the symptoms (I thought), but it never developed, so I guess I couldn't claim the Purple Heart. My toes didn't have any feeling in them at any time, numb and perhaps sometimes tingling. But I finally attributed this to the cold weather and found out it was a common ailment.

The time for moving came, and found us not quite finished with the last touches on our trucks and equipment, but that could be taken care of later on, maybe. The last night in Delta Basin was spent in carefully packing our own personal gear and getting things cleaned up for the trip. We all had the problem of scientifically cramming the duffel bag with all the stuff that was presumably required for our living overseas, cramming it so that nothing poked out. This was a problem, but finally we solved it much as we had back in Myles Standish and on board ship: brute force in mashing and crushing everything to get it in. We wrote our last letters for a while—by candlelight in the tent. We had to write from beneath the covers, with our hands sticking out freezing. Life's little hardships.

At length everything was in readiness and we were given small, four-section mimeographed maps and directions for the whole convoy northward. It described cities, refueling points, water points, bivouacs, emergency medical service and ordnance and motor repair service. It listed travel control points (TCPs) and route numbers and all the details so that we couldn't get lost. (But, of course, our truck got lost just the same.) We studied the maps for a while, and then bade goodbye to the sunny Riviera.

On the cool, cloudy morning of February 1st, we lined the trucks up on the road just north of the company area. We'd gotten up early and taken down our tents, made up our packs, stowed them in the trucks, eaten and then warmed up for the take-off. We sat around in the trucks waiting for the signal to move, and made last minute checks of truck, equipment and gear, just to be doing something. Jeeps with officers kept going back and forth getting the convoy in line.

I was assigned as official driver of one of the weapons carriers. Riding with me was my temporary team chief, a T/4 named Wally Masters. Wally was a heavy-set, somewhat nervous chap with thick glasses, and he kept worrying about everything being all right. The company clerk also rode with us, another T/4 named Ben Wilson. He elected to ride in the back, where it was warm and soft on the duffel bags. He could see well enough by bracing his chin on the loudspeaker and looking out through the front with us.

Finally, the time came and we began to move, slowly at first, then picking up speed as we moved out of Delta Basin. Speed stayed at around 35 miles per hour and we were given instructions to stay about 60 yards behind the vehicle in front—a dispersed convoy, tactical in case of air strafing attack. This, we hoped, wouldn't be likely because of the shredded condition of the Luftwaffe at this stage in the war. Nevertheless, as we rolled on through the countryside, I kept my respectable 60 yards behind.

Everything went well at first. We passed some beautiful country—rolling hills with rocks outcropping as small cliffs. There were farmhouses and small villages and there seemed to be grass growing and trees with leaves on them, though how this came to be in the cold of January and February we didn't even try to figure out. The morning turned into afternoon and I was thoroughly enjoying this little sight-seeing trip as we drove kilometer after kilometer deeper into France, ever northward.

I wasn't getting in the least tired, but Wally Masters began to insist that I must be, and that he should take over the driving and relieve me. No, I said, that wouldn't be at all necessary because it was such easy driving.

About midafternoon we stopped for rest room purposes and I got to talking with the radio sergeant, Lister. I happened to mention that Wally had wanted to drive and Lister told me it would be best somehow to refuse diplomatically. Wally was nervous, he said, and besides, with those thick glasses he might not be able to see as well as others. So I resolved that I was going to be boss and drive the whole trip.

Shortly before dinner we drove into the town of Avignon, a historic place that had been the continent's papal center during the time of the great schism of the Dark Ages. Apparently we enjoyed the history, the scenery, and the pretty girls too much, for we got lost.

Consulting the map, we found Route N-8 to Aix, N-7 to Avignon, Orange and points north. So we led the entire convoy behind us through a circuitous route in the streets of Avignon and finally surrendered ourselves to ask a native in broken French how to get to Montelimar. A few minutes later we rejoined the trucks up front, sheepishly following along where we ought to be.

We passed around the western edge of some high, rocky mountains that we assumed to be the French extension of the Alps. The roads in France we found were narrow. French automobiles were about half the size of our American ones, and the foreign trucks not nearly the size of our tremendous dreadnoughts in the USA. So to put a convoy of American Army trucks over this road required some careful driving and tight squeezes. What made it bad was that, probably for the sake of beauty, every road in France seemed to be lined with trees that had been placed a foot or two off the pavement. This led to the feeling of driving down a country lane but a number of the trees had their bark barked by the bumpers of passing vehicles. We passed one Jeep that had plunged into a ditch. All this led to what you might call precision driving.

Wally never let up in his insistence in spelling me at the wheel. Since he was a T/4 and I only a Corporal, he felt he could insist on driving and get away with it. He kept on until he arrived at the verge of pulling his rank. He ordered me to let him drive. I should never have caved in, but I've always tried to be somewhat diplomatic where others are not. Like a fool, I agreed to let him drive. So we changed at the next rest stop and he took over.

Then, 12 kilometers north of Avignon, it happened.

Those narrow French roads and most unyielding French trees were the indirect cause and part of the end result. We were humming along smoothly down the country lane, minding our own business. Wilson was in the back looking out front and talking with us. I was in the right front seat with a box of .45 slugs on my lap loading the clip of my M-3 sub-machine gun.

Suddenly, around a curve we approached a civilian truck coming toward us and, at the same time, a civilian bicyclist on our side of the road. Wally's problem was to keep on his side of the road so as to pass the truck and yet get over toward the middle so as to miss the man on the bicycle. Fortunately, the truck came first, and Wally nervously got way over on the right to avoid it. Then, quick as he could after the truck got successfully out of the way, he wheeled with all his might right out into the middle of the road to miss the bicycle. From my point of view, I was sure we were going to crush the civilian, bicycle and all, but we didn't—merely grazed him and gave him a gentle push that set him off balance so that he had to stop. We, however, didn't fare quite that well.

"Wow! That was a fast stop!"

After Wally had gotten turned out into the highway to miss the civilian, he didn't quite know what to do. We were moving about 20mph and heading for a tree on the opposite side of the road. He got a half-Nelson on the wheel and began to turn with all his might back into the road. But there is one thing about a $^3/_4$-ton weapons carrier—it will not turn as sharply as you'd like it to turn, at least not on a dime. And somehow, Wally never thought of putting on the brake.

Result: ***CRASH!***

Now, at 20 mph, you might think that it couldn't have been much of a wreck at all. Only the left fender hit, and only the left wheel was rammed underneath the motor. But it had a series of results. I braced for the impact and only banged my head lightly on the windshield and scarred my knuckles when we hit. After the dust cleared, I reached over and turned off the ignition and looked at Eddie. There he was, slumped down in the seat, probably out cold from shock. I spoke to him a couple of times and he mumbled something about his leg.

I knew that it would have been next to impossible to injure a leg in this little smack-up, but I looked down and confirmed it by seeing nothing amiss—at least no broken bones. I touched one leg and, eyes still closed, Wally said it was the other one. I touched the other leg and he let out a sizeable yelp, deciding then that he would come to.

I forgot about Wally and turned back to see what had become of Ben Wilson. He apparently had been worst hurt, for he had made the mistake of watching the wreck with his chin near the radio loudspeaker. When we hit, he was dashed against this and received a large cut under the chin. He wasn't aware of it, and nearly fell over when I told him what had happened to him. He let out a yelp and stretched across to look into the mirror, let out another yelp and crawled out the back to go seek a first-aid man or medic.

Wally crawled out of his side of the truck moaning in pain that I was sure he only suspected he felt. I looked down to see if I had been decapitated but found only that .45 slugs were scattered all over the seat, the floor boards and the ground outside. I got out. The Frenchman stood by the road with his bicycle, scared to death not from a close shave but perhaps the thought that we might execute him for having interfered with a troop movement. Another Frenchman came up, offering advice and help. I told him in French that no one was hurt and everything was all right, so he retired to the other side of the road with his countryman.

Meanwhile, the trucks in front of and behind us had stopped and our radio pals were coming forward to see if they could assist in burying somebody. Wally tried to find a broken leg but never had any success. The only thing to do now was wait for Ordnance to come along with its wrecker. So I began to pick up .45 shells and continue the loading process from which I had been so unceremoniously interrupted. Then I found time to notice that I had been fairly well cut across the knuckles by not having sense enough to keep my hand in a safe place during the mishap and had received a couple of scratches on my lower leg. By now, a goose-egg had begun to grow on my forehead.

The three of us decided to split up and continue the journey in other vehicles. So, crying over the spilt milk that I wouldn't get to do any more driving up to the front, I crawled in with Luke in Smitty's truck ahead and the trip resumed. Our duffel we left in the old truck but we didn't leave the scene of the accident until Ordnance had arrived with the wrecker and taken the injured vehicle in tow. Then, on to war.

"N7 to Dijon, left on N8, right on N9, N217 to Lyon. That clear?"

TWENTY FOUR
CONVOY

The disadvantage of riding in the back end of a truck was that you could only see where you'd been, but we saw a lot this way, just the same. We passed German trucks and tanks that were twisted, burned-out piles of wreckage, now covered with rust. We passed villages and saw a number of advertising signs, though none more frequent than "Dubo... Dubon... Dubonnet," apparently singing the praises of some choice wine. We passed magnificent chateaux, with superbly designed architecture and landscaping—some sitting high on the hillside at the base of a tall cliff, others sitting close to the road.

We drove through rain for a while, then stopped for dinner in a large field where the trucks could be assembled. Wally,

Ben and I circulated around showing off our wounds, and the fellows ate it up. We, of course, were the first battle casualties and that was some distinction. By now my goose-egg had increased to its maximum size, and Wilson was going around barely able to talk with the big patch that covered his chin and half his face. Ah, it's wonderful to be a war hero…

The convoy followed route N-7 from Avignon through Orange, Montelimar, Valence, St. Rambert, Vienne and on to Lyon, following the Rhône River much of the way. At Lyon we went by the grounds of the Université and saw nattily dressed young students strolling along on the campus. They had rapidly reverted back to peace time, it appeared. At Lyon, too, after driving over the big bridge we considered to be across the Rhône, we passed a block where a bazaar was in progress and the brilliant colors on display there smarted the eyes.

We had not seen sunlight for several days and on this first day of travel up from Marseilles the landscape had seemed a dull gray, brown or green. That's why we enjoyed the difference by observing the gaily colored cloth of bright dresses and scarves and table cloths, along with hundreds of other items of fashion and household. The bazaar was busy, too, for people seemed to be milling everywhere. Naturally, the passing GIs saw some "babes," as they called them, that would be worth milling with. Soon the bazaar disappeared down the way, and we settled back as the trucks pulled away from the city and rolled north again, back into *la campagne.*

At Villefranche we hit the junction of Route N-6 and followed this to Chagny, where we picked up N-74 to Dijon. We picked up something else along in here. Snow. Lots of it. Looked like about 12 inches had recently fallen. By the time we reached Dijon, night had set in so we refueled the trucks and pulled into a bivouac area. There was little but C rations to eat, but by now we were hungry enough to admit that they tasted pretty good.

The only place to sleep was in the trucks. In order to avoid overcrowding, Wally, Wilson and I gravitated back to our wreck, which we found after considerable and diligent plowing around in the snow and mud. We had to have one of the Ordnance boys lower the front end so we could sleep in a prone position, or at least somewhat level. Finally, we made up our beds, though I found that it would be pure fallacy to suppose I could sleep in the truck bed—which measured a scant five feet in length. I therefore selected the bench at the side of the truck so that when I wanted to stretch, I could push my feet out the back end into the cold. After doing this once during the night I decided I'd rather be cramped than to have frostbite. Even in such a close and precarious position I got some sleep, believe it or not, and never once fell off the bench.

Breakfast was filet of Ration C, au gratin, and we were off again. We continued northward from Dijon along Route N-74 to a point shortly past Langres. Here we turned off the main convoy route and hit Routes N-417 and N-460 past Meuse, Bourbonne, and Darney to Epinal. This last was a fairly good-size *ville*; we arrived there in late afternoon and were told that this was to be the bivouac for the night. We unloaded our blankets and chow and went off to a building not far from the parking area—a building that would serve as our bedroom during the coming night.

It turned out to be quite an extensive boudoir… a flour mill several stories high. We explored it from stem to stern, looking over the abandoned flour bins, elevator shafts, and other decaying sections. We did find life down in the basement, where a small bakery was in use by some Frenchmen. They were turning our rich, brown loaves of exquisite appearance and tantalizing aroma. We stayed there and talked to them as long as we could stand it without raiding the place, then went back upstairs, found a cozy corner, swept away some debris, and lay down to a night of dreams on a flour sack. *Voilà! Nous nous souviendrons Epinal…*

Breakfast the next morning was filet of Ration 10-in-one, en casserole, a la Française. This was our introduction to the famous 10-in-ones, and we liked them. There seemed to be more of a meal in this type of eating, and I particularly enjoyed the cereal that came with it. Just pour in water and serve, and it really tasted good. We also had our little Coleman stoves that would heat an individual cup of coffee at the same time. Hot meals without a mess hall. Nothing but style for us…

Next day, we headed north from Epinal and deeper into the ETO, closer and closer to the front. Evidence that we were approaching the hot spots showed up in some of the villages we passed—where lovely French girls came out to the trucks and talked to us rather than flee in panic. They had been so newly liberated that they hadn't yet had time to tire of the American soldier, and his newness with them was still an adventure. Once in a while the convoy was somehow held up, and when we stopped, up came the populace. Once we had a kissing bee with

"Gil does get carried away, doesn't he?"

the people of a French house. Another time, the boys threw out a pack of cigarettes and got a tantalizing bottle of wine in return. Then, too, everything that was in ruins here began to appear as if it had been freshly smashed. We had no idea where we were going and as far as that goes, we didn't have much knowledge as to how things were shaping up in the war in general. We'd been out of contact with the news of late. This may have sounded silly with us being radio men, but we were given instructions that it would be best not to use the radios, so we maintained a communications silence.

The end of the convoy for us was Morhange, way up in north France close to the border—how close we didn't know. We didn't particularly care to know how close we were to the front lines either, but they told us. Mighty close, brother, mighty close. So we went into tactical thinking and tactical acting. It didn't help our nerves any to discover that Morhange was only a brief stopping place, a bivouac, as it were. On the morrow sometime, we would be shifted up into the combat zone. As to Morhange itself, we couldn't make out much in the rain. We drove into what was obviously a former military barracks—a group of large red brick buildings arranged in a square, with a large space in the center that held a few trees. We piled out and were assigned a building in which to make our quarters for the night.

Shortly after getting my gear from the damaged truck and piling it in one of the rooms, I was informed that I would be on guard duty that night. "Oh, no!" I screamed to myself. "What if there should be an enemy push, what if snipers should infiltrate the lines and sneak back here to the rear echelon, what if…"

This, my first night at the front—guard duty! Exposed. Vulnerable. Doomed.

I was too tall for this, too young to die, but I resigned myself to it and took consolation in the fact that my insurance premium had been paid up to date.

Before twilight disappeared, I had a brief look around the building. We were right, it had been a military barracks for the German army. It was in ruin now, but there were still swastikas all over the place. There was no furniture of any description; all the rooms were barren. Over one entranceway I noticed a placard that read: "Gedenke das du ein un Deutsch bist!" Freely translated, that meant: "Remember that you are a German!" I hoped they *would* remember it and stay on their own territory during the night. I strolled back to my room and thought about cleaning it up a little and spreading my sack so that it could be hit with ease when I got off guard duty just before dawn. I found some semblance of a broom and swept up the debris. Conveniently, in one corner there was a hole in the floor, so all the debris went down through the hole.

Night came. And so did guard duty. I tested my M-3 to see if it was loaded to the hilt. Thirty rounds, I felt, could do a lot of damage before they got me. The gun had been cleaned and oiled and would fire at my slightest command. I gave slight little commands to myself all night to fire at those ungodly noises I kept thinking I heard, but somehow the hair trigger never went off.

My imagination filled the darkness with morose and sullen Nazi troops whom the war had bypassed. Incensed that these American invaders had wiped out their dreams for a brave new world—*Deutschland über Alles*—they had come to pick off someone in revenge. And who better than a guard at night? They could sneak up on him, pull out their Lugers, aim, press the triggers, and **POW!! Zi-nnn-ng! Whack!**

It was dark as the ace of spades. Here this close to the front, there was total blackout, so I only had the little noises for company, and since I exaggerated all the noises, I had a lot of company. My particular area was around part of the building and included a big drinking bag. This made it convenient. Every time my throat got dry from having an imaginary close brush with the enemy, I just stepped up and got a drink.

There was another fellow on guard whose duty was to patrol the other side of the building. We'd like to have stayed together for the purpose of more adequate protection when the enemy came but we couldn't. We would meet in the dark, often mistaking the other for a sniper, greet with the password, and then talk for a while in low, whispered tones. We found plenty to say, but finally had to break it up and go off into the unknown to make the rounds of our post. I felt better after these encounters. He was scared, too. With this, I reinforced what I had learned as an actor on the stage. Enduring Lesson Number Five: *When you are scared to death, act nonchalant.*

ENDURING LESSON #5

"When you are scared to death, act nonchalant."

The night was not without its movement, though I'd rather there hadn't been this type of movement. One of the buildings down the square was being used apparently as a field hospital. All night long, ambulances came in from the front lines—swinging in through the gate and around the square, visible only by the narrow slit of their blacked-out headlights. The ambulances backed up to a door and then there was a little more light as medics brought the bodies out and moved them inside. At least I thought they were bodies. Actually, it was only the wounded being brought in for more careful attention from aid stations closer to the fighting. Just the same, I could see myself on one of those lovely stretchers, dying for my country. In spite of all the macabre imagination, it was an interesting night and wasn't too badly fraught with tenseness. The real tenseness we saved for the next day.

Relieved of duty, I got a little catnap during the dawn hours and then was awakened to prepare for the move up to the front. I packed my blankets and duffel together and stowed them in a truck to which Dody Ratigan and I were assigned. After a little preliminary instruction, we left in a small convoy and started north.

It was a cloudy, drizzly morning and anyone with any sense wouldn't be out in it. Least of all, enemy planes. Nevertheless, we could fancy a lone Messerschmitt zooming over the horizon to riddle our truck with holes. We proceeded with caution, observant and tense. Dody drove while I held my M-3 in my lap and looked carefully in all directions, prepared to meet the enemy and have it out with them. But not a single thing happened on the trip, much to our relief, and we decided to quit acting scared.

The war was far from won, but we were not yet in immediate danger. We drove in to St. Jean-Rohrbach that day, our first front-line CP (Command Post). Here we were to go into combat at last, to test out the training we had been months in getting, to use it now in the last drive to defeat the Hun invaders. It wasn't, however, as glamorous as it might have been. We were vulnerable, and we were in the thick of things, but sometimes when the action was not very close by, we referred to ourselves as garritroopers: too far back to git shot, too far forward to wear ties.

TWENTY FIVE
SAINT JEAN-ROHRBACH

Once again we found ourselves back with the boys of the 70th Division. We had joined the Infantry now, to work with our own, our very own Trailblazers in blazing a path through Lorraine and across Deutschland. The 250th Signal Company had been serving with the regiments and we pulled in to replace them completely. Now we knew where we were and what we were doing. Though we couldn't write it home, we were near the northern border of France, fighting with the Seventh Army, XXI Corps, under General Patch.

St. Jean-Rohrbach itself was small, a typical village in the Lorraine part of Alsace-Lorraine. It sat on a broad knoll, spreading on both sides of the main highway that passed on north. There weren't many houses, even fewer stores and hotels. The main street was scarcely more than two blocks long. The greatest feature of Rohrbach was its mud, for this was February and winter was in the process of turning into spring. It was still cold, especially in the mornings, and we had some sleet a couple of times, but mostly there was mud… deep, sticky mud everywhere except right in the middle of the paved streets. When you stepped into the most innocent-looking brown puddle, it turned out to be a quagmire about a foot deep and you almost left your shoe there in getting the foot out. We still had our shoe pacs and they kept the mud out.

Our radio section was quartered in part of a bombed-out building that had obviously once been a hotel. The rooms were small. Four of us were assigned to one and though you couldn't cuss a cat in it without getting fur in your mouth, we threw in four bunks and called it home. There was a little stove and a table, and with this not much room for us. We made it cozy though, and wrote during evenings by candlelight. For fuel, we managed to find some coal, a little wood and cowchips, anything that would give off fire.

The top floor of the building was devoted to rooms for the section chiefs (BTOs, or Big Time Operators, as we called them) and a radio room. This area saw most of the action. The walls of the building were a foot and a half thick, made to withstand the assault of time and erosion, but not of blockbuster bombs or artillery barrages. Yet the month we spent with this as our base headquarters was a pleasant one to a great extent. Far better than the bivouac at Delta Basin. We weren't always in the region of home base, but when we were, it wasn't a bad place to come home to. We had a lot of bull sessions, laughing, and good times there. And by the way, there was a slit trench outside that served as a latrine, but no one used it. Going around behind the barn out back seemed somehow easier if not more romantic.

The mess hall was located in a building just to the north from and across a muddy field from us. If we were hungry we went across the muddy field. If there was plenty of time, we went around. Usually, we were hungry. But Norman the cook and his boys put out plenty of food and it was mighty nigh good all the time. We also got daily PX rations of cigarettes, a Hershey's Tropical chocolate bar, stick of gum and, from time to time, shaving equipment and other sundries. I had a weakness for the chocolate bars, especially after they were heated and dehydrated on the top of our stove. When we tired of them, we stored them up and then roasted a batch later for an afternoon snack.

The cooks had a big job and we each got KP a couple times while we were overseas and ate like pigs. Later, after we'd been in combat for a while, Jerry PW's took over the KP jobs and we had life easier. They were glad to take over that particular job, too, for it meant something for them to eat for a change.

Everything didn't always move like clockwork. We had our combat sleeping bags, the kind which zipped up and had the advantage that if they were zipped all the way up through the top, the bag fell open. This meant that if shells came popping our way or the enemy came rushing in, we wouldn't be trapped in the bag. Just rip it up and jump out. However, we had another reason for getting out of the bag in a hurry. There came an epidemic of dysentery. When we got the GIs, it was amazing how fast we could move and the sleeping bags were ripped more than once. One time one of the boys got caught and didn't get out fast enough. Well, those are the fortunes of war.

As spring began to approach northern France, the cold was interspersed with nice warm sunny days. In St. Jean-Rohrbach, this brought about something else. It so happens that a principal crop of this farm community consists of beets. To store them for the winter, the farmers buried them beneath piles of straw and manure in the front yards of houses. During winter, most of these piles of manure remained frozen. When spring came, this material began to thaw out and the more liquid elements of it ran down the sides of the streets, forcing us to watch our step carefully when circulating through the village. With all this came that fragrance common to manure piles that melt in front yards. At times we thought of Rohrbach as a wide open *pissoir*.

Meanwhile, the war went on, and we were in combat radio. The BTOs on the upper floors took care of operating the Division communication nets for the first week or so and I could never understand why they were always so nervous. Some of them would come down to us with their hands almost shaking and say that they had just sent their first combat message.

I asked them: "What are you afraid of?"

"Well, think man!" was the commonest reply. "What if you made a mistake? You might send hundreds of men to their death."

I frowned. "Well, for crying out loud! Don't make mistakes!" I must have sounded more like an officer than a non-com; at any rate, I seemed to have been bereft of politesse. "Haven't you had enough training under tactical conditions back in the US to make this duck soup? Don't you know your code? You've got messages to receive and messages to send. Why is this so different that you got to get excited and scared? The whole bunch of you act like sissies!"

If they didn't throw me out on my ear, they came back with: "Bullshit! Just wait till you get up there and send your first message. This is the real thing, kid..."

I gave up. Maybe it was because I'd had the lead in stage plays in high school and had become accustomed to acting calm under any circumstances. I expected them to do the same.

When they finally got enough of my braggadocio, they marched me upstairs and sat me down on the key of a fast Division net and stood aside to smirk. They expected cold sweat to break out and my fingers to start trembling so that I couldn't even move the key. But me & Benning & McClellan—we sure as hell fooled them.

All at once a station called in (Morse code, of course) to say he had a message for me. The BTOs scrambled for paper and pencil so that at least one of them would get the message. They were convinced I'd mess it up. I gave the station the go-ahead signal and the operator began sending. There was nothing in the room then but the *dits* and *dahs* of code and the furious scratching of pencils…

At length, the message was finished and one of the boys, in a panic, said: "Ask him to repeat the last ten groups! Quick!"

"Why should I?" I retorted, with as much sarcasm as I could muster. "I got the whole message perfect!"

I sent the station a Roger and the sergeants sat back in their seats with a look of astonishment.

"From now on," I said to them, savoring the moment, "don't ask me to repeat something you didn't get. And I'd advise you to keep your trap shut out of courtesy to the man who is supposed to be officially at the key. And now…" I gave them a smile with a trace of a sneer, "what in the name of hell is there to get so excited about on your first message? Really, if you can't take code you might just as well transfer to the Infantry and put your talents to better use."

This was pretty abrupt talk to superiors, but I thought they could take my genial tone. Not so. It made them hopping mad. Section chief Moriat, a cigar-chewing hulk of a master sergeant, said he was sure I'd messed up the message, so we got the M-209 converter out and started to decode it. It came out perfect. Then Moriat looked at the decoded message and said: "Ah, HAH! I thought so! Look, the Zs didn't print!"

This bears a bit of explanation. The M-209 converter is so constructed that when a message is being deciphered the Zs never print. This is the only way in which a space can be put between words—by recording a Z. Sometimes, though, some words like zero, which have a Z in them, will not print right, yet you can make sense of the message if you know that a Z is missing.

A little further explanation is needed here, not about the convertor or the Zs, but about the man who made the stupid conclusion that the Zs didn't print. That was typical of Moriat.

Moriat was unquestionably the most bitterly despised man in the outfit. As the highest ranking non-commissioned officer in the radio section, he should have been able to outshine most of us in code, yet he couldn't even take 16 words per minute. They said he'd been to the Signal Corps school in Fort Monmouth, N.J. and had flunked out. What he knew about procedure in radio operation could have been printed on a postage stamp in headline letters. In a word, he was the worst man in the section, and how he'd gotten to the position of section leader I don't think any of us knew.

Everyone laughed inwardly at his stupidity in radio and no one respected him. He gave it to us new kids from Benning, stick and twist. We were on details as fast as he could think them up, and we seldom got a moment's rest. He made all of us mad because he would never allow us to have a bull session in which we laughed. I know this is hard to believe, but I'm giving it to you straight. When he heard us having a good time, he would come downstairs and find something for everyone to do... even the assistant section chief Lister, who could never get along with him. Oh, once in a while, perhaps, he would join in the fun, but he was such a loudmouth we always looked for a way to break it up.

One night, Moriat gave me the axe, but good. It was about midnight and I was deep in the soundest sleep I think I'd ever experienced. All of a sudden he burst into the room yelling at the top of his lungs: **"SUTTON! SUTTON! GET UP QUICK! GET UP! GET UP!"** He started shaking me.

Every wheel in my head began to spin the wrong way, and I tried like the devil to shake the heavy sleep out of my head, but it wouldn't seem to go. I wondered what was up—if the enemy had us surrounded or we were being bombed. I sat up slowly, still struggling to open my eyes to see in the dark and straining to get my wits together.

"GET UP! HURRY! OUT OF THE SACK!" He yelled a few more times and then dashed out of the room.

I fumbled clumsily around to get my boots on and get them laced. He raced back in cussing and yelling for me to hurry.

I was still straining my eyes to see as I followed him outside to the radio truck.

"Hop in!" he yelled, jumping into the passenger's seat. "Let's get going."

I got the truck started and turned into the driveway that led from our building out to the main street. The road was in such a muddy condition that there were only two narrow high places upon which the wheels might travel. Getting off these would mean probably sliding off and getting stuck in the ditch. To make matters worse, the headlights were blacked out by slitted covers so that only a faint light shone ahead to illuminate the road. I leaned forward over the wheel and tried desperately to squint and focus my still cloudy eyes on the dim path ahead and was suddenly hit by a severe pain in the eyes and head. This apparently resulted from the shock of hitting the deck out of a sound sleep and straining the eyes to see under the worst possible conditions.

We drove across the main highway to another building that housed the Division Command Post and spent an hour setting up and getting into communications with an Artillery net. Moriat kept trying to put in his two cents' worth by telling me how to tune up the set and get the right frequency, things about which he knew next to nothing.

We made it, and got the messages over, and at dawn I closed down the station and went back to our own building. But I kept that headache for three days. After this, I joined wholeheartedly with the rest of them in a bitter hatred for Moriat. Something would have to be done about him, and, we thought, the sooner the better.

During our first week in Rohrbach, three of us were assigned to go out with a battalion of the 275th Infantry and operate a station for them. We made a night ride up front to a little town by the name of Gubenhausen, a name German enough to convince us we were right close to the border. This was the closest we had yet approached to the front line. We could hear rifle and machine-gun fire not far away and the rumble of artillery fire seemed to be all around us.

Off on the horizon, we saw flashes of fire that indicated the direction of the battleground. We stowed our truck away in the barn of a private house and camouflaged it. We stayed there a couple of days and made friends with the owners of the house. They were good hosts and made our visit a pleasant one wherever they could. And we gave them cigarettes and one thing and another to express our gratitude. This was, to them, quite a contrast to the actions of the German Army that had so recently vacated their village.

Getting out on advance stations like this suited us just fine, for it took us away from the CP and Moriat. We hoped for more such situations where we could be our own bosses. Then, too, it made fraternizing so much easier. We liked the people. We wanted to meet as many of them as we could.

TWENTY SIX
THE CENSORS

Throughout February, the mud continued. It was especially heavy in the area where we had our motor pool and our trucks would sometimes have to move around in mud up to the running boards. There was mud on the highways, too... that is to say, there was a yard-wide muddy patch on the roads between the edge of the pavement and the trees that lined the way. Since the highways were too narrow for our big trucks to pass, they had to go off the roadway into the mud where they could, sway and swerve a while and then get back on solid pavement.

I had a driving complex that lasted several weeks after the wreck we'd had on the way from Marseilles. I had an inner, unreasonable fear that we were going to hit something again. There was no way to control it. It was just there and I had to contend with it. I didn't do much driving because of this, but I did have to ride in trucks driven by others, and some of the drivers were daredevils of the first degree. I took one ride in the back end of a $2\frac{1}{2}$-ton truck out to one of the nearby villages one day and after watching the muddy road zip by at about fifty miles an hour I began to say my final prayers and ask forgiveness for all my sins. We passed some other trucks and went swaying off the road just barely missing those beautiful trees. Having once hit one of those beautiful trees, I didn't care to have the experience again, so I just closed my eyes and tried to think of something else, something safe—like being up in the front lines. Eventually this thing wore off and I became a full-fledged driver once again.

An interesting aspect of life in Rohrbach was working the huge 399 radio set. This was the one housed in the big truck. It was entirely covered, had an air conditioning unit, tables on which to write, and cushioned seats. It was linked to the XXI

Corps net as the 70th Infantry Division station and the operators here functioned with more efficiency and snap than did those in the lower echelon Division net. The 399 operated day and night and its generator consumed a great number of five-gallon cans of gasoline every day.

After we got to Rohrbach, the bottleneck of mail broke and we all got an armful of letters and packages. One of the most pleasant things I could find to do when in the room or on duty with the radio sets was to take out a little packet of cocoa from home and brew a steaming hot cup, as much for the delicacy of it as for something to do to keep awake. Of course, every room and every radio control room had its own little stove and pot of coffee.

And we often did have to fight to keep awake, for the flow of radio traffic on any busy net had a mesmerizing effect. When a *blippity-bleep-bleep* tone of code, transmitted automatically from some distant station, repeats itself in the background of your earphones, there is a small line between waking and dozing. Dozing was unthinkable in combat, so we used every possible means to stimulate our senses and keep our brains functioning. Besides cups of coffee and cocoa, we used small bits of food to munch on, and we read our mail countless times.

Some letters were slow to catch up and came at the most ironical times. I still carried on a mock feud with my friend in the Navy who had gone into service before me but still hadn't gotten out of training. "You lucky dog!" he wrote. "Already you've seen five times as many states as I have, but just remember you've never been out of the U.S. and I have been to Mexico!" I had a good laugh out of this. Later he went to midshipmen's school at Notre Dame for higher training. It still wasn't overseas, though, and he champed at the bit. Naturally, I gave him the prongs in every letter, saying I was over here in France, had seen some fabulous scenery and met some fabulous people, and was fighting and dying for my country and keeping the Germans where they couldn't come over and hurt him. With that, the inter-service fur really flew…

After so long a time we began to miss the little things we'd taken so much for granted back home. For example, we had obviously kissed goodbye to the delicacy of ice cream, and we hadn't seen a movie in such a long time we couldn't remember back. But we resigned ourselves that it would be a long time before we dipped our spoon into a chocolate sundae or watched Betty Grable prance across the screen. We kept right on requesting boxes from friends and these kept us in some contact with civilization. Women—American women—were also quite rare. There were some WACs, some nurses, some Red Cross girls, but the officers found more free time to visit them and occupy their time than we did and so took them over.

Once, though, a lovely Red Cross girl set up a doughnut-making machine in part of a building in Rohrbach, and I assisted her one day in making a batch of heavenly sinkers. I also assisted her in getting rid of them, for I ate thirteen of them before I knew what I was doing. Enduring Lesson Number Six: *When you find a lot of doughnuts, eat 'em.*

We talked for a long while and I would have liked to talk for a longer while, but others started to sample her culinary wares and I kept busy at the machine—still gorging. That's the way it was with the little things; sometimes the most insignificant items turn up as an unforgettable luxury overseas.

When going out on field operations, we made use of emergency rations. We had the 5-in-one variety (enough rations to supply five men for one day) and these were good. They had that delicious pre-mixed breakfast cereal, corned beef, peas, ham & eggs, cookies (more affectionately known as dog biscuits), butter & jam, cigarettes & candy, and lemonade or coffee mixtures. Even

"Happens every time the Red Cross brings doughnuts."

the C rations we got sometimes were good, especially when heated. There were occasions upon which we would wish for more emergencies, so as to get the rations. But our regular chow wasn't bad, so we didn't have much complaint.

Another day we had the sun with us and afterward noticed that an almost imperceptible part of the French mud had begun to dry up. Hints of spring appeared, and we felt we'd soon be able to step out of our winter cloaks and take a breath of fresh air without coughing. However, we selected one of the icy cold days that followed this particular period to have our first shower in the ETO [European Theater of Operations]. Almost none of us had been out of our underwear since landing in Europe and it felt good to expose ourselves to fresh air once again, even though the underwear did walk right out of the door. We stood our pants up in the corner and went in to the shower room. This was an Army unit, a portable shower installation with canvas partitions. It worked and we felt wonderful to be clean again. Then a shave, and fresh, dry clothes, and we felt as if we could lick the whole world.

Licking Germany was being done for us by the 70th Division Infantry. Came the news reports:

> *"From the 7th Army front, Stars and Stripes Correspondent Ed Lawrence reported that the Germans were still offering determined resistance last night in a forest west of Forbach, after the town was cleared by the 70th Infantry Division. Other units of the 70th pressed on to clear the southern part of the Styring-Wendel, where a prisoner of war stockade was surrounded and overrun. A total of 1,250 Allied captives—French, Russian, Czechs and Polish—were released. Four hundred could not be moved immediately because of illness, it was reported, while 600 were suffering from tuberculosis."*

Other combat stories spread around the Division. A German soldier was the last man in his company to get home on a Nazi version of rotation, but while home on furlough he heard that the Russians had encircled his old company. That was enough. Loading up all the rations he could scrounge, he headed away from the Russians and promptly ran right into the PW cage of the 276th Infantry. The *Stars and Stripes* also reported this one:

Doughboys of the 70th Infantry Division are getting suspicious of anyone in female attire. A number of Germans captured in Forbach today were found to have slips and dresses under their uniforms. If the fighting got hot, apparently the Krauts were all set to strip off their uniforms and pose as civilians. But the Division men have worked out a solution. They look at the feet of all suspects. The disguised Jerries are sure to be wearing boots, since women's shoes are rationed in Germany.

One fault of our section radio officer, a First Lieutenant named Sandifer, was that he never thought it important enough to tell us what and what not to write home in our letters. From time to time we would come upon taboo items that we'd already put into our letters thinking they'd been innocent enough to get by. We got to the point where we didn't know what to write, so one day I put a stop to this. I wrote the following V-mail home:

It is quite beyond me why I sit down and think that I am going to say anything other than that I am in good health. There seems to be nothing more to say that would pass censorship here. You have probably had the opportunity to read other letters from overseas and perceived that therein was enclosed much more information about that individual's activities than I have been able to divulge during the entirety of my visit to France.

I know it sounds extremely stupid for me to say: watch the papers for news. I can't write anything. The situation is that we can't even write home a lot of stuff that's spread through eight pages of our STARS & STRIPES newspaper here. We just aren't told what and what not to write! That may explain the condition that some of my past letters have arrived in.

Otherwise, all is pretty nearly well. It was somewhat cold today and even tried to snow, but Spring in Europe is insistent and the buds are popping out on the trees. Hoping that this innocent little missive isn't another of those chopped-up, inked-out masterpieces you know so well, I am...

Next day, Sandifer called the whole section in and referred to a letter that had been written by one of the members of the group regarding censorship. So we spent an hour getting thoroughly oriented on the current rules governing censorship. After that, letter-writing was a much easier task. Enduring Lesson Number Seven: *When things get all screwed up, write a letter to some one.*

While we were in St. Jean-Rohrbach, there began a detail that took us frequently away from the Division CP. Though we did go out on radio work occasionally, it was work with telephone and telegraph wire that kept us busiest. At the time we were at this place, the front had remained relatively stationary, with sporadic fighting centering on this bank of the Saar River. Since the Division wasn't moving, most of the communications network was laid down with wire—a lot of wire. Much of this wire had been put up in the hurrying days of capture and since that time, other nets had been established, making the old wire useless. Now it wouldn't have been nice to have left that old wire lying all around the villages we went through, so the only thing left was to police it up. So my teammates and I went out on a number of wire-policing details to outlying villages, loosening the wire from fences, poles, buildings and trees, and winding it up on drums in the truck. Some may have thought this unromantic detail, and a helluva way to win the war, but I found it had certain hidden advantages. It gave an excellent opportunity to get acquainted with the people of the border country of Northern France, to see how they lived, how the war affected them, and what they thought of us. Moreover, I began to learn the language to the point where I was becoming a secondary interpreter for the company. And there was plenty of occasion to use this broken command of French. N'est-ce pas?

ENDURING LESSON #7

"When things get all screwed up, write a letter to some one."

TWENTY SEVEN
THE FRENCH

We saw a lot of magnificent country on our wire-pickup work, and we saw a lot of devastated country, too. We got in to one place near Bening where we could hear fighting in the forest across from us—sporadic bursts from rifle and machine gun. But this wasn't as menacing as the rear echelon groups who went out into the fields for target practice.

On one occasion we almost got hit this way, and had to come up with a white flag of surrender to indicate we were friendly troops. Though there were pleasant green fields and forests, there were also the bombed villages, the stores and hotels and homes that had been blasted by artillery, the twisted rails and littered wreckage of trains, the ruined bridges—vast areas in shambles. We found bomb craters in the road, in the fields and on vacant village lots. Patches of ground here and there had been marked in several languages: **DANGER, MINES!**

"What did you say, George?"

Our wire work took us to Wustweiler, where a young French maiden had us up to her room to talk to her because she was lonesome. Seems her boyfriend had gone off to the war and hadn't been heard from, so we talked to her, that is, I interpreted some for the boys... whatever I wanted them to hear. We could have talked about love or any other such desirable subject and the boys would never had known.

We also got to Loupershouse, where one of the men visited a certain house all afternoon. He took in a pound or two of coffee in order to make his way and apparently succeeded. At Loupershouse, we talked to a bride-to-be and she told us of all the gay plans they were making for the coming nuptials. She even showed us her gown.

At Loupershouse, I thought I would be smart and learn pole-climbing, a process necessary in extricating wire from high places. So I put on the spikes and climbed clear up to the top of the 20-foot pole, gaily loosening the wire and letting it fall. Then I looked down and almost fell off. The ground looked 50 feet away, and I learned what they meant when they said climbing up a pole was much easier than going down. The pole shook as I came down.

I made it all right, and bravely climbed another. This time it wasn't the same. I started down, got about eight feet from the bottom and slipped. After sliding down and landing with a jar that rocked my teeth and spine, I examined myself for splinters but luckily the pole had been a fairly new one. That, however, happened to be the last pole I ever climbed. Six-feet-five, I said to myself, doesn't do it.

Our wire work took us to Metzing, Farebersviller, Seingbouse, and Puttelange—all partly destroyed by war. We found ourselves out on all-day excursions that were interspersed with trading for food. A typical bargaining dialogue went like this:

"Pardon, madame, et bonjour."

"Bonjour."

"Avez-vous du pain et du confiture ici?"

"Ah, non monsieur, je le regrette..."

"Pour cigarettes?"

"Ah-h, oui! Oui! Un moment, s'il vous plâit."

We almost always got our loaf of French bread and a little jam to go with it. One lady invited us in for a sauerkraut dinner a la homemade. Another baked a cake for us, that is, we bargained for a cake and brought all the materials, but it turned out more like an apple pie. It was delicious just the same.

Our foraging around for food wasn't entirely limited to the French people. We once came upon the mess hall of a company of Negro truck drivers at a time when biscuits and butter and jam were just waiting for the gulping. The cook invited us to eat as much as we could so we spoiled our appetite with hot biscuits, one after another. We found this cook to have the best philosophy of any cook we had ever encountered in the Army. "The boys drive hard and work hard all day long," he said. "They come in hungry and they want something to eat. Something good. It's my job to give them something good every meal, and that's what I try my best to do. The best soldier is the well-fed soldier, so I feed the boys like kings."

From the taste of the biscuits we devoured, he wasn't kidding in the least. They invited us for supper, but we figured we'd better get back to the CP before the MPs came after us as deserters, so we thanked him and told him we had to leave. But we sure would have liked to stay around there for a while.

Radio work took us one evening to a house in Diebling. Here we found a rather well-educated couple in their fifties and combined our food with theirs for a delicious meal. We talked about a great many things, including mostly the war. There were a number of words we would have to act out to get the idea across, but we learned a lot of French ideas that night concerning a torn-up world. They seemed convinced that the war wouldn't last more than a couple of months longer and we hoped they knew what they were talking about it. They should. They lived here. Their name was Lomme and the man thought it would help pass the evening if we should engage in some game of wit. He suggested one called *moulin*, but I had never heard of it, so we thought of something else. He mentioned chess.

Ah, I said to myself, this is just up my alley. Hadn't I been in the High School chess club and become an expert? Well, this guy would be a cinch to beat, so I might as well show him that American superiority in the game. We got the board out and he beat me in five minutes. We played another game, with the same result. I was much shaken, but I guess not really surprised. These Frenchmen were exceptionally skilled at the game. And why shouldn't they be? I'd forgotten that chess had originated over here in Europe. Nevertheless, Monsieur Lomme and I became fast friends that night, though I never saw him again.

As to the impression of the people in general, I had this to say at the time: "This is France, after practically four years of war and occupation, but the people seem to be doing not too badly. They have adequate clothing and while some are not clothed too well, many of them are attired with a semblance of Paris finesse. It's nothing to see a cart drawn by one cow and one horse, and oxen are abundant as beasts of burden. It's rather a quaint sight to observe an old man and an old woman diligently plowing a hillside field—all day long, regardless of weather. They're workers here and they keep their houses as clean as a pin—mud or not. Their main trouble is lack of food and cigarettes, so I guess they do have a pretty rough time of it."

Such, then, was the land into which we came as the liberating Army. We weren't the ones who were hugged, kissed, offered wine and showered with flowers. That was for the doughs who first came through driving the Nazis out. In many of the towns, one of the main streets would be called AdolphHitlerstrasse, but these signs were torn down in short order and the street suddenly renamed.

Sometimes, with a few moments to spare, I would sit down, take out a pencil, and jot down the episodes that transpired, impressions on the spot…

THE OLD LADY OF FARSCHWILLER

"Bonjour, Madame."

"Bonjour, Monsieur."

"Avez-vous du confiture ici?"

"Ah, non, monsieur." Her eyes were steady from behind the thick lens.

"Oh, I understand."

"Non," she looked downward, at her broom, at the soft, damp earth.

"No sugar?" I asked.

"Oui. Pas d'sucre. The Boches took it all with them."

"Yes, I know. I thought… I thought, peut-être…"

"I am sorry, Monsieur. There has been very little."

"Savez-vous ou puis-je obtenir du confiture? Perhaps the neighbors would have some."

"It is the same everywhere."

"Oh."

"When the Boches came here to Farschwiller they took everything from us. There is little now. They also took—."

She broke off. Behind her graying head there was the tidy, barren porch, and the dark interior of the patched hovel. She was looking down again, now—but not at the broom, or her shapeless dress, or the earth…

"I am deeply sorry, Madame."

"Ma petite fille. They have her. I do not have her now, she is gone. I—I have not heard from her. She was west with them. But now—I do not know."

She looked up and shook her tired head.

"Votre fille?" I asked.

"Oui, monsieur. My little girl. She was almost twenty, monsieur, and then… then, they came. I do not know. I have not heard of her."

"Je comprends, Madame. Merci beaucoup."

"Yes," said the others, "thank you for your kindness."

"Service, messieurs. Au revoir."

"Au revoir—." There was nothing else to say. That was the way it was. At the next house the door screamed slightly as it opened.

"Bonjour, Madame."

"Bonjour, Messieurs."

"Avez-vous du confiture ici?"

❦ ❦ ❦ ❦ ❦

THE LADDER AND THE APPLES

The land between Cochern and Bening was pleasant: two similar grassy vales separated by a small ridge so that one *was lower than the other. There was grass everywhere, in the fields that had not been plowed, up the railroad embankment, and sometimes between the twisted rails; the woods cast an austere feeling of fertile grandeur about the countryside.*

From the ruined bridge to the junction there lay a pastoral scene of simple, exquisite beauty—behind, the hills with timber patchwork, ahead, the twisting road that struggled up the slope. Bening itself reposed in an interlude between the rich valleys; the church up the mountainside, the duck pens in the valley, and the village itself squeezed in between—that was it. And the children playing in the street, who burst into song with the arrival of the strangers. Their song was from a land far away, the words unknown to them, yet they sang with little distortion.

"You are my sunshine, my only sunshine.

"You make me happy, when skies are gray.

"You'll never know dear, how much I love you,

"So please don't take my sunshine away..."

"Madame, uh—. Avez-vous un ladder?"

"Qu-est-ce que c'est cela, Monsieur?"

"Why, uh . .. c'est un... Oh, how do you say—. Avez-vous une echelle. Vous comprenez ça? Une echelle?"

"Ah, oui, oui! Venez, venez-ici!"

In the barn, she and her aged husband struggled several moments in extracting the ladder from the hay loft.

"Voici, Monsieur."

"Okay, where do you want it?"

"Let's get it under that window up there."

"There, now. But wow! How it shakes. Here goes."

"Hey, bud, better come back down. It's not strong enough to hold you."

"OK, you talked me out of it. But how are we..."

The Madame approached and offered aid.

"Oh, oui. We've got to get that wire loose, see—. Regardez. Fils de ferre."

"Ah, oui, oui, Monsieur."

Without hesitating she flung herself up the weaving ladder and was soon reaching out to free the tangled tele-

phone line. It was, however, too far out from the wall. The window opened and the old man appeared with a long broom. This he pushed out and began working with the wire, his wife directing his labors with a series of cackles.

Below, the strangers watched and saw, after a bit, the wire come loose and fall to the ground with a slap. The Madame slipped down the ladder, her feet working with the same agility as in the ascent.

"Oh, Madame, thank you, very much. Merci, beaucoup. Vous êtes une grande dame."

Whereupon she became filled with glee, which measurably increased when she was proffered a pack of the stranger's cigarettes. Her rapid thanks became profuse.

"For your husband," he said. Still chattering and looking after him, the Madame hoisted the ladder into the barn.

The strangers continued about their duties to the corner, following the wire line. A small lad, appearing distorted and disfigured at first glance, but actually frail and tiny, skipped to the corner and proceeded to play for them on a miniature accordion. The tune was harmonious and carried its fast-moving melody cleverly throughout the octaves. It was quite probably some border air that had been sung again and again through the years.

It impressed the strangers.

Before it was finished, the Madame of the ladder came running up to the one who had presented her with cigarettes and placed into his hands four small and wrinkled apples.

"For you," she said, "and for your friends."

"Oh, merci, Madame. Merci, beaucoup."

She backed away, having offered her gift, still exuding enthusiastic appreciation. The stranger distributed the apples—quite possibly a goodly share of the household's stored foodstuffs—and they were tasty. Before the afternoon had swept away, the Madame had presented the strangers with another collection of apples, and she thanked them for their kindness. They thanked her for her help and for her apples.

As the sun set behind the haze that hung over the mountains, the strangers departed from Bening—drove down the winding road to the junction, then along the grassy vales and toward Cochern, where the land was pleasant.

❦ ❦ ❦ ❦ ❦

TWENTY-EIGHT
MERLENBACH

We soon got used to the rumble of the artillery and seeing its flashes of light after dark. Once in a while there would be a period of quiet and then the blasts would cut loose again, sounding like three thunderstorms going at once, then stop after a short period only to take up the usual noises again. It was this sudden outburst of the big guns that at first sent us rocketing skyward, but then, when we became combat-hardened soldiers, we took it in stride.

In the latter part of March the Trailblazers saw considerable action—not the big push, that didn't come until later, but the boys up front were kept busy just the same. We didn't always get the straight dope and the report that one whole company of one of the Infantry regiments had been wiped out was probably just another war story the Garritroopers promoted to have something to do. The 70th had by now come up to the Saar River and was taking up positions along that natural barrier, while cleaning out pockets of resistance and consolidating forces.

Stories kept coming in from the front. They told of how a single artillery shell had prevented demolition of an important bridge across the Saar. It came from a German PW, an officer whose mission it had been to destroy the Altbruck, the main highway bridge across the Saar. Six hundred pounds of TNT had been placed in each pier and wires run to the explosive from an electric detonator. On the night of March 16th the Germans decided the 70th was on the verge of an attack against the Siegfried forts along the river, and the bridge was ordered down. The detonator handle went down—but nothing happened. Frantic investigation showed that a shellburst a few moments earlier had severed the wires. Accurate time bursts sprayed a hail of shrapnel which prevented a repair crew from reaching the break. The Germans did succeed in blowing out one of the middle sections but this was quickly repaired by the 70th Engineers when the right time came.

They told of a combat team from Company E, 274th Regiment, that was clearing out a schoolhouse in Styring-Wendel. In the cellar they discovered a rifle of enemy make. They were holding it when footsteps approached. It was the Kraut coming back looking for his rifle. When he saw where it was, he decided the safest course would be to surrender. Which is exactly what he did.

This little tidbit appeared in our Division newspaper:

There wasn't a straddle-trench so Staff/Sergeant Eugene Stouder, Company C., 275th, dug himself a cat-hole and was just using it when he saw three men disturbing his privacy. Not only that, they were Krauts, full grown and combat serviceable.

Noticing that they seemed much more curious than warlike, he finished his rites, hitched up his pants, then motioned them to follow. And led them back to his foxhole where he picked up his carbine and took them back to the company CP in the proper manner. Now, Mr. Anthony, his problem is this: How in hell can he tell the story of this heroic feat to his girl friend back home?

One interesting thing came out of German thinking when they began to capture the Krauts right and left. They were ready, most of them, to accept the fact that America and Russia were winning the battle of Germany. But one of the biggest hopes they had, and one that was pounded home relentlessly by the Nazi propaganda machine, is that Russian communism and American democracy would not be able to mix in the post-war world, and that these two countries would fight among themselves. We smiled at this thought at the time. Why, how ridiculous! The Russians, our friends, our allies, our comrades! They wouldn't think of any such thing. Actually, it's just too bad that someone doesn't have a crystal ball in which he can look and predict with full accuracy the things to come and help to preserve the peace of the world. But then, that's neither here nor there, and we didn't give it any thought then. We were busy with World War II and would let Number III come later if it must...

During the first week in March, we moved from our Headquarters in St. Jean-Rohrbach and set up housekeeping in a town much nearer the German border—a place called Merlenbach-Freymingen. It was two cities in one, I suppose, but we cut it down to just plain Merlenbach. When we got there, the bright flashes of artillery fire just beyond the hill to the north of the town were big and close and we knew then that we'd been brought up mighty near the action.

Merlenbach was a mining town in the Saar Basin, and you could easily tell that its chief product was coal. Even though the mines weren't much in operation at this particular time, there was that telltale smudge over the buildings of the town that had been brought down from all the smokestacks for centuries. It was a pretty good-sized town, though not at all a city, and had been surprisingly little damaged by the war.

For our Signal quarters we occupied an old two-story schoolhouse. Here, we figured, had been taught the warped and twisted ideologies that brought us over here in the first place. We ate chow on the little desks and wrote letters on them by candlelight at night, blacked out the windows, rolled our sleeping bags on the cold hard floor and called it home. For a little while.

After circulating around for some time, we found out that even though this was still in France, it was so close to the border as to be German for every practical respect. Names and language were German, yet here, as in all border towns, both languages were commonly spoken, probably depending upon ancestral lineage which was to be the preferred one. And in the couple weeks we visited Merlenbach, a lot of things happened.

The whole company moved into the schoolhouse, lock, stock and barrel, and turned it from a building of desolation into a beehive of activities. Living quarters and some offices occupied the second floor; mess hall and dining room on the ground floor; motor pool in a concentration on the large playground outside. We didn't have much radio communication from this point because wire lines had been thoroughly installed. Some of us spent a lot of time with USAFI (U.S. Armed Forces Institute) courses and some nights found me deep in some intriguing algebra problem. Then there was sightseeing around town and fraternizing—the last allowable until we swept into Germany.

One night, for some reason or other I don't call to mind, a T/4 named Pete Smith and myself decided not to wash our mess kits in the normal mess hall water set aside for that purpose. Perhaps it was because for once in the ETO we wanted our eating materials cleaned up good, and we ourselves didn't have all the time in the world to become good dish-washers. So we said to ourselves: Supposing we took some soap out to one of these houses around the school block here and see if one of the ladies might clean them for us. The French were exhaustingly clean about everything they did (as far as we knew). So we called at the nearest house, 51 Rue Minor, and knocked at the door.

This little effort began an extremely pleasant association that lasted throughout our brief stay in Merlenbach. We introduced ourselves to the family there and they invited us in. We got our mess kits washed and a lot more besides.

Their name was Crochovalsky. Polish by descent, they had come to France years ago and in the process of all this transferring and living near the border there was a decided confusion of languages in the family. We negotiated principally with the younger son, Conrad, a handsome lad of about 22. He spoke five languages: the Polish of his native land, the German of the border country, the French of the country on the edge of which he now lived, Latin he had studied in school, and English he had also studied academically.

His parents spoke Polish and a little smattering of French, so they and we had an almost impossible time. Smitty and I decided that here was quite an appropriate place to learn languages. Smitty selected German, I stuck to French. We learned fast, too, and wherever we had trouble, Conrad would dip painfully into his meager English recollections and come forth with the saving word. It worked well that way and we had pleasant evenings. Of course, we kept them well supplied with cigarettes and candy and what other luxuries we could spare or come upon at the time.

In return, they treated us as best they could under the circumstances. Wine, of course, was brought out the first night.

And I immediately made the mistake of saying I didn't like wine. This, to them, was impossible—for they had never heard of such a thing—and I must definitely be joking so here, have a glass. I had a glass. Protests were in vain, and I soon quit for I got the impression I was hurting their feelings by saying I didn't want any. I learned a valuable diplomatic lesson those nights, Enduring Lesson Number Eight: *If you must gulp down a foreigner's favorite drink, try not to retch until you get outside.*

ENDURING LESSON # 8

"If you must gulp down a foreigner's favorite drink, try not to retch until you get outside."

Worse yet, they were so generous that they didn't just fill a little wine glass for me—they chose a full-sized kitchen drinking glass. I winced a time or two as it went down, but they didn't see it. And once, when I wasn't looking, they filled the glass clear to the brim.

Sometimes I wondered if it's worth it to keep on being a teetotaler, and yet trying to drink wine or beer or what-have-you is, to me, like pushing down castor oil. I'd just as soon have a glass of that yellow cocktail that ran from the manure piles down the streets of St. Jean-Rohrbach. Unfortunately, the Crochovalskys not withstanding, and all my fake sipping notwithstanding, it was a physiological case. I could be nothing else but an outcast. But a friendly outcast, of course. They pitied me.

Tiring of our solid floor bunks at the school house, Smitty and I spent a night with the Crochovalskys and found what it was like once again to sleep in a normal, full-sized bed. It was the first time for me since Springfield, Missouri, back in that distant land of home. Next day being Sunday—and we for once had a Sunday off—we went to Church with the family. The churches in most of Europe had been spared by the advancing Allies and had seen little damage. The church in Merlenbach was not an exception to this though one or two stray artillery shells had hit it; and the holes had been repaired. It was a beautiful place, with magnificent stained-glass windows and ornate archways and finish. The morning sun came through in ethereal streamers of varying shades, casting a celestial glow over the entire inner chamber. Services were held both in French and in German, at different times of the morning. We went to the German services at 11:00 and though I didn't understand a word, it was nonetheless impressive. Almost all of France, it seemed, was Catholic and I felt somewhat out of place since I did not belong to that faith. It was interesting to observe the bright finery in which the people were clothed: gaily colored dresses and hats, fine, thin sashes, and other elegant decorations—truly a contrast to the dreary colors of war.

By the time our visits with the Crochovalskys ended, we learned a lot of foreign language. Smitty was fortunate enough to have his camera with him plus a means of developing film, so on the last day of our visits we took a series of pictures of the whole family: mama, papa, the three boys and even of the neighbor's family. Smitty later mailed them prints. Conrad gave me a ring made out of a pre-war deux-franc piece, which went into my souvenir box. I should very much have liked to have put Conrad's sister-in-law (Benjamin's wife) in my souvenir box, too, but then, looting was prohibited by SHAEF (Supreme Headquarters, Allied Expeditionary Forces). This was the friendship struck up by knocking on a door and saying simply: "Voulez-vous laver deux assiettes?"

In Merlenbach we had places to shower. Down from the hill on which the schoolhouse overlooked the town, across the footbridge that crossed a railroad yard, up the steps and into the mine buildings. This was one of the largest showering establishments I had ever witnessed. Coming out of the mines, of course, meant for the miners a shower each day before going home, so here in a vast building were located 300 showers. The miners' belongings were kept away from sticky fingers by an unusual though simple device. When each miner went on duty, he went along the hall in the shower room to a place where he removed his street clothes and donned work attire. Then, attaching his good clothes to a hook, he pulled them up to the roof some twenty feet above by a long chain and pulley, then locked the lower end. Nobody could high-jump up to latch on to his possessions that way. When not working, his mining clothes hung safely up in the loft.

On the approach to the main shower room there was a sign hanging over the hallway that read, in English:

FREE SHOWERS
COMPLIMENTS OF THE FRENCH MINERS,
BUT THEY HAVE NO SOAP

While we showered, we occasionally saw some Frenchman working slowly up and down the line, looking in the wall brackets for soap that we might have left behind. He found some, for we always left plenty, usually taking a new bar and leaving it after one bath. I talked to one of the miners and asked him all about the mine. He said that although for obvious reasons they were operating on a reduced scale at the time, ordinarily this was one of the biggest mining areas in all Lorraine and the Saar. But he personally was soon going to retire. What would he do when he retired? I asked. He said he would spend most of his time pursuing his hobby. I asked him what that hobby was. His reply was simple and quick: his wife.

It was in Merlenbach that disaster first struck the 570th Signal Company. It began one morning about one o'clock. Everyone was snoring in deep sleep on the floor in the bedroom, when section chief Moriat came in and began waking some of the guys. I waked as he moved around from sack to sack and wondered sleepily what was up that required night operation. Was this, perhaps, the big push?

I lay there quietly and listened. He waked up a couple of the boys and told them to get up and get dressed and go out to warm up their truck—a night radio patrol was in the offing up toward the front.

I turned green with envy there in the darkness. Those lucky stiffs getting called out for a combat mission, and I fervently hoped Moriat would rouse me out of the sack and let me go with them. I gave that up, knowing how he and I felt toward each other. So he and the boys finally went out of the room and everything once again became quiet. I turned over and went back to sleep.

Next morning, I'd forgotten about it until the whole company was shocked by the news that came in mid-morning.

Two trucks had gone out. One was driven by Randall Foley, and he and his crew got back. The other truck had Apley as driver, Darmer as gun guard, and Sergeants Lister and Masters (my old nemesis in the truck wreck) in the back with the radio set.

The trucks went up toward the front lines and, driving by blackout lights, got lost. It seemed that Foley's truck had taken all the wrong turns that Apley had taken before him, so Foley, driving behind, saw the whole thing. They drove along roads clear up to the front without knowing where they were. At the boundary of American forces, the Infantry had placed a road block to prevent Germans from sneaking vehicles across during the night… a road block that meant that the road had been heavily mined. The guards fired above Apley's truck to warm him to stop and not run into the mines. Apley and Darmer mistook these shots for enemy fire and gunned the radio truck right into the mines.

Apley died within an hour, Darmer in three days. Masters and Lister were thrown out of the back and suffered less. Masters received a foot injury that eventually got him back to the States, while Lister, after receiving several weeks treatment at base hospitals, later rejoined the company.

We got the story next morning, and there wasn't much laughter around the school house that day. The thought accidentally occurred to me that my wish might have come true when Moriat was circulating among the sacks last night: I might indeed have been chosen as a driver. And Mother might have shortly been putting up a Gold Star in the window at home.

A bitter irony came out that day in a rumor that began to circulate and later picked up considerable confirmation: the night patrol that had been so costly appeared to have been merely a practice patrol instigated by the assistant division signal officer. Up to that time, we had only small respect for the officer in question; and after that time we had no respect. That day the Message Center boys went out to the scene of the crime and quickly picked up their secret converter M-209. But they didn't pick up much more than a hunk of melted metal. Mines in roads will do that sort of thing.

Another night in Merlenbach was filled with excitement and everything but sleep. This time it was about midnight or a little after and we were all slumbering peacefully in our sacks, or at least as peacefully as you can on a hard floor. I know I was deep in slumber and sawing redwood logs. All of a sudden, the most terrific explosions began to take place one after another and we began bouncing around and shaking as if the entire school building was coming down around our ears.

It seemed as if all of France were being picked up and crumbled over Germany. The noise was deafening, and it kept up. Now when you are awakened out of a sound sleep by such explosions as these in a combat zone, you are apt to begin wondering all sorts of things. And I did. Shells, I thought, but then came that horrible, haunting question: Were they *incoming* or *outgoing*?

It took what seemed like an extremely long time to figure out the answer. Since I hadn't yet been blasted out of my sack and the school house was still intact, I decided they were outgoing. And they were. It seemed that a whole battery of big guns had been moved into the school yard on the east side, and the artillery had set up for action. When the proper time came (and certainly without warning to us) they cut loose with all the guns at once and with all they had.

We got up and looked out the windows and there they were, huge flashes of light and thunderous explosions rocking the earth, as the guns lobbed shells across the hill and into the enemy lines. We began to wonder what was up, and in a couple of days we found out. This, at last, was the big push. We were off for Berlin. Better hide, Adolph, your days are numbered now.

Crossing the Saar on a pontoon bridge.

TWENTY NINE
DEUTSCHLAND

We didn't leave Merlenbach directly but were kept busy. They picked a most inopportune time for Marlene Dietrich to put on a show for the boys; she had a show in the afternoon and I was planning to go see her in the evening, but that was cancelled for lack of attendance. So I stopped thinking about lovely film stars and went back to winning the war. The way we were going, it certainly did look like we would win it, and in a hurry.

My radio team went out on a night ride to operate radio communications for one of the Infantry battalions across the Saar and up near the thick of things, though combat now wasn't too much more than chasing the Krauts home. We crossed the Saar River just a few hours after the Engineers had gotten their pontoon bridge across it. As a matter of fact, ours was one of the first trucks and they watched us carefully all the way across—guinea pigs, I assume, to see just how well their bridge was holding up. Just like crossing the Big Piney on a rickety bridge. Their faces had a worried look, but you should have seen the look on my face. I wasn't particularly anxious for a bath in the Saar.

We went through Saarbrucken at dusk and it is impossible to describe the scene of utter desolation there. Saarbrucken had once been a good-sized city, now it was one big and vast badland of twisted wreckage and piles of brick. There didn't seem to be any living thing here, no people, no light, no fire, nothing. That was what 70th artillery had done to it, plus the work of B-17s and other bombing missions. They compared it with Aachen, Cassino, Cologne.

Strictly tactical, we drove in the darkness using only the slits of blackout lights. Beyond Saarbrucken we entered a little town called Fischbach. We pulled into a country yard and set up our radio by remote control inside the house and operated all night long. One message we handled turned out to be about the longest we experienced throughout our tour of the ETO: 82 groups. A steady dose of Morse code.

This, however, was just a one-night stand, so we returned to Merlenbach next day and prepared to shove off with the rest of the company and enter Germany officially. As we drove toward the Saar River, which marked the border, we went through the town of Forbach, which had been so bitterly contested by our Infantry. We passed a sign of welcome from the boys of the 70th Division:

Photo courtesy Harley Rush.

And then we were in Deutschland.

We soon found out what it is like to be in on a big push like this, when the Americans struck out for Berlin and there was no stopping them. After a while we figured there would be no stopping us, for it appeared the ride was never to end. We wondered how we could be so deep in enemy territory, on enemy home ground, and still remain alive.

We had no idea where we were going or how long it was going to take to get there, and if it hadn't been for the MPs all along the way, giving us directions and route numbers, I think we'd still be lost. We kept going through village after village, and sometimes alongside long lines of vehicles. Eventually we came to an intersection and were given orders to split up.

Our team, now fairly permanently composed of T/4 Joe Kowski, a San Francisco boy, and T/5 Mike Morrow, a Brooklyn lad who knew German well, was assigned to proceed without delay to a place called Landstuhl. Somehow, through all that maze of roads and in the tactical darkness of night in Germany, we arrived in Landstuhl in the wee hours of the morning. There was a moon shining and that, fortunately, gave us some light to work with and by which we might get our antenna up and the radio installed for immediate use.

One of the first things we noticed was that there were ubiquitous white flags of surrender—in trees, on roofs, up flagpoles, atop barns, churches, everywhere. Obviously the Landstuhlians desired that their village be spared the onslaught. This was new. This place had just been overrun, and we overran it some more. We set up our communications in the Town Hall. There were some Germans running around trying to help us and we gave them orders now and then just to let them know who their new bosses were. We were still green at the art of occupying a newly taken village and we still had with us a hatred for the Krauts, any and all. They didn't trust us, either, probably suspicioning that we would execute them if they didn't bow down and tow the mark to us.

Of course, we took the long way around and did it the hard way—we built a flattop. This is a particular kind of antenna installation that is useful in getting out long distances with maximum communication and is of considerable usefulness in hilly or mountains country where natural obstacles make radio communications difficult. The flattop was hooked to the roof of the Town Hall and to the top of a tree out in the yard. While Joe was getting the set together in one of the front rooms of the building, Mike climbed the roof and I turned monkey to get up in the tree. It turned out to be a much bigger tree than it had looked from the yard below, but after some ripped clothes and a few scratches, we got the antenna up and were at once in communications.

While Joe was on the radio set and Mike taking a snooze I wandered around the offices of the building and saw evidence of sudden evacuation everywhere. Perhaps it had been occupied by a Nazi regime in Landstuhl or had been used as an observation post, but whatever the use, it had not been left in a neat condition that should be characteristic of City Halls. I found different kinds of documents all over the place, none of which I understood, so gave it up and went down to take my shift at the set.

We didn't see much of Landstuhl for this turned out to be only another one-night stand. Next morning, we received orders to pack up. We got our antenna down, removed our remote set-up, and rejoined the Company to continue the ride deep into Germany. That day we crossed the Siegfried Line at or near Homburg and later went through Kaiserslautern, a large southern Germany city that had not received the devastating attack that had laid so many other places to ground level. And then at last, in late evening, we arrived at our new quarters—Otterberg.

THIRTY
OTTERBERG

Here we were in a beautiful part of Germany where there were rolling hills, picturesque mountains, and spruce-fir forests of exquisite beauty. Otterberg (*mit Abteikirche des chemaligen Histerzien serklosters, Bauzeit 1190-1225*, which means abbey church of the former Zisterzienser's Monastery built between 1190 and 1225) was itself a quaint and picturesque mountain village. From our vantage point we saw a large church building in the shape of a cross, with high arched windows.

We also saw what appeared to be a school or a hospital, and numerous close set houses with sharp-pointed roofs punctured by garret windows. Far in the distance, we could see rolling hills covered with patches of forest and a checkerboard pattern of farmland. There were castles on some of the higher mountains, and these lent a historical air to the countryside.

Up in the forest near Otterberg there was a Neues Schwimmbad, an exquisitely beautiful swimming pool, though we didn't get to see much of this because the weather hadn't yet warmed up. Spring, however, was on the way, for showers had opened up and there was considerable fog during our stay in this village. I watched a flock of sheep near the village one day, with some of the cutest young ones you ever saw, and I don't see how they kept from catching cold since they had just been sheared and looked like outcasts from a sheep nudist colony.

We lost no love on the Germans; there were still the white flags of surrender and the streets named AdolphHitlerstrasse. But we couldn't feel continuously bitter toward the average citizens. It didn't particularly please us to have to evict the family from the nice house that sat on the hill above Otterberg. I went up with Mike, who served as interpreter, and Lt. Sandifer, to tell the people they must vacate for a while. They were an elderly couple (we found out later that he was a dentist by profession) and they were startled at the thought of having to leave. Startled, confused, and hurt. But we figured that if any of their countrymen had ever come to our country and evicted our people they would have done it at point of bayonet, probably much as they had done over in France. Perhaps these people had relatives that had been in the German Army and had killed American soldiers. With these thoughts in mind, it didn't pain us too much to tell them we'd be back in an hour to take over. Besides, the couple should have no trouble in finding someone to live with. There must have been a number of vacancies in the village.

With the passage of an hour, we returned and moved in. It was for us an extreme paradise. There was a large, beautiful, exquisitely furnished living room with a view window that looked out upon the village below, framed by fir trees in the yard outside. This was one big benefit of being a radio operator. We needed high elevations, so we always got the best views. Behind the house an orchard sprawled out over a grassy slope. Joe and Mike and I selected upstairs bedrooms for our private quarters. I found my bedroom furnished with one of the softest beds I'd yet come in contact with. Besides that, there was a large feather tick, like an overgrown pillow of luxurious softness, that covered the bed and was the only cover needed.

Here we reveled as the better-than-garrison soldiers who never had it so good. There was even a radio with broadcast band, mahogany cabinets in the living room, and a couple of typewriters, or "schreibmachinen," which made letter writing much easier and faster, and we took turns using them. This was something. Maybe war wouldn't be so bad after all. Let the Krauts pay back some of the discomfort they had caused us.

News came in and we found out what was going on and what we had to do with it. We had somehow been transferred into Patton's Third Army, but we had been relieved of active combat and were now to settle down to the job of occupying the vast territory the Third Army was sweeping through. Patton had made a brilliant move. With his drive toward the Rhine and Frankfurt, he trapped a number of German divisions, as well as American GIs who were fighting these divisions. By the time this pocket of resistance had been cleaned out, Patton's boys were in Frankfurt and the front was a hundred miles away. In warfare, a hundred miles is a long distance.

So the war had left us. We set up radio control in the basement of the house on the hill and rigged up a comfortable place to work. The basement had other things, too. There was a converted laundry, there were bins of potatoes and fresh and canned cherries—which, of course, we sampled. There was plenty to eat, including boxes from home, and once in a while we would interrupt ourselves long enough to answer the phone... "Wyandot 28, Cpl. Sutton... No, she was a brunette... Milly... Yes, that's right, the redhead we just called Red, and she loved it... Me and Mike... Asleep... Yeah... No, no business at this end... Good! Keep the home fires burning, it's going to be a l-o-n-g war... Oh, you bet! See you at chow..."

And to think that we were getting paid by the day.

We found three motorcycles in one of the nearby sheds and got a little practice riding them, but not far. We tried to find room in some of the trucks to take them along, but the authorities frowned on theft. One of our radiomen once found an outfit of ultra-violet light communications equipment that some German outfit had left behind, and we went up on the top of the hill behind the house to try it out one night. The apparatus consisted of a tripod on which was mounted the sending light with which to transmit messages and a pair of black light binoculars so that light from the other station could be seen. Power was supplied by a wet cell. So we got lined up, centered our binoculars on each other's beam and began transmitting.

It worked well and we carried on quite a conversation, talking on invisible light. This was the type of equipment, apparently, that could only be used from CP to CP, and to outposts, only if those points were located within sight of each other and probably not more than a mile away. Then it would have been an ideal means of communication that the enemy might never intercept, for the stations could not have been discovered in the first place.

It was at Otterberg that we finally got enough of Moriat. He had been pulling his stunts long enough and beating us into the ground by the blunt end to the point where we got really fed up with him. All of us complained to Sergeants Lister and

Morton. That wasn't actually necessary because they were just as sick of him as we were. So one day the top stripers in the section called Lt. Sandifer in and had a private conference with him in one of the rooms of the house. And that did it. Moriat was ousted for good from our little group.

Next day happened to be a day for the reading of the Articles of War to the entire company, so while everyone was gone, Moriat came to the house and got his gear. I watched him from the basement window where I was on duty in the control room. He didn't seem particularly perturbed for perhaps he expected something like this.

It also happened that our Supply Sergeant had just been investigating a house in Saarbrucken that he shouldn't have been investigating and got blown up with a booby trap. He wasn't killed, and for all I know he's still in the hospital. Therefore, a vacancy existed in that position and Moriat filled it. Oddly enough, from then on he was nice as pie to the radio boys. Whenever we went to the supply room to draw some item of equipment he would go out of his way to give us the best in stock. Oh, well, they say it takes all kinds to make the world. However, the less of his kind, the better the world.

On the 7th of April, I changed from winter longies to Spring underwear. Those flannels had never been any too desirable anyway. They'd never reached to my ankles and after the first washing they shrunk up to the knees. My back kept getting cold, too. Thus it was distinctly a relief to change into something more comfortable. I soon gave up trying to keep my trouser legs in my combat boots; getting 34 x 36 trousers overseas was about as easy as getting a Coke.

The German schreibmaschinen were fascinating gadgets. They had numerous extra characters and the whole machine had so many knobs and levers that the simplest operation required considerable concentration and cut-and-try. Besides that, the y and the z were reversed, which certainlz made tzping less easz.

We got PX rations semi-monthly. They consisted of candy bars, beer, cigarettes, cream wafers and peanuts, a great deal less than the packages that arrived in the mail from time to time, but at least we knew some of the delicacies were coming all the way down the line. We got plenty of cigarettes so that we could commonly carry around a boxful in case of some black-market emergency. If we ran out, we visited the boys who didn't smoke and they would supply us, that is, if they hadn't just been out on a foraging expedition themselves.

Some of our spare time was spent in cleaning the M-3s, but this wasn't as difficult a task as with the M-1s. The stub-nosed M-3 sub-machine gun was small and compact, well fitted for drivers. We sometimes wondered if we'd ever get the opportunity to use them. The M-3 fired at the rate of 300 rounds per minute, although there were only 15 slugs in the magazine. It wasn't much on accuracy, but it sure could mow 'em down. We had to fire them once in a while to make sure they were still working. But that's the life of a radioman—no excitement, and it's the penalty for being a garritrooper.

We sat around and took life easy. There was some rain and drizzle and fog at Otterberg but it wasn't accompanied by cold. At least, after Marseilles, I don't call anything above zero cold. We sat in the front room and had our bull sessions and listened to the radios. French radios, by the way, probably looted from France and brought back by the retreating conquerors.

While we were so enjoying life here at Otterberg, we knew we'd better get the most out of it because, as with all good things in the Army, it wasn't destined to last long. Thus did we revel in the luxury of a soft bed, beautiful view, and all the comforts of home. And yet the curious thing about it was that the next place we got to was even better than Otterberg. That may sound impossible, but lend an ear and I shall speak of many things… things of beauty, things of spice, and things that aren't like Army life.

THIRTY-ONE
BAD KREUZNACH

During the first week in April, we moved on deeper into Germany and landed at the beautiful resort town of Bad Kreuznach, on the Nahe River. We learned that the prefix of "Bad" on the name of any German town signified "bath" and meant a health resort boasting usually of mineral waters. Bad Kreuznach was one such place, and believe me, for us, it was distinctly a health resort, too—though we never visited the "badhaus." We stayed in the Hotel Haus Kipper and couldn't have wanted a finer barracks.

Joe, Mike and I selected a corner room on the second floor, a room with folding French doors that opened out on a small balcony where we could sun ourselves and look down at the gardens and breathe deeply of the invigorating scent of lilacs in full bloom. In the hotel yard and along the street grew grass and hedge, and there were trees everywhere. This we liked, and there was more.

The room we selected had a beautifully carpeted floor, indirect lighting from an orange-colored chandelier, black polished table with cushioned chairs to match, cabinets equipped with drinks and reading material, a writing desk of polished mahogany, end tables around a soft velvet lounge, a magazine rack adorned with three potted plants, long curtains and draperies, and an exquisite tapestry on the wall. Adjoining room: dining. Across the hall: bathroom with hot and cold running water.

You may find this hard to believe. We did. And you may find it hard to believe that there were some *disadvantages* to this heavenly setup. (1) We had to sleep on a cot (not enough beds to go around). (2) No maid services. (3) Only 24 hours in a day.

It was apparent that the people living in these apartments had been evicted suddenly for almost everything had been left untouched. Drawers and cabinets were full of personal equipment, closets hung with suits and dresses, linen cabinets well supplied.

One day the owners came up and asked to be admitted. With full non-fraternization policies still in effect, I answered: "Nein!"

"Bitte?"

"Nein."

"Zie kommen mit mir?"

"Nein!"

The soft soap didn't work, so they started to leave, pausing by the cellar door and looking back plaintively.

"Nein!" I repeated. "Du kann nicht gehen!"

They kept on looking and I repeated: "Nein."

"Ah-h-h. Alles nein!"

"Ja, ja… alles nein."

This was mean and I was merciless with the Krauts. I didn't really feel any sorrow for them, nor pity, even if they had been the victims of twisted leaders. For I continually compared the luxurious homes they had here with the hovels of the French, and took careful note of how few of these villages had been bombed out. They had not been bitterly contested by artillery fire since the military advance had been rapid here, so the war had scarcely touched Bad Kreuznach. On the other hand, German soldiers had taken 20-year-old daughters from their mothers in France… That dwelt on my nerves. I wasn't made to be mean, and I didn't like being mean to people. It was just one of the heart-wrenching conflicts of war that I had to accept and put up with.

One night, sectional ingenuity brought forth a chicken dinner complete with mashed potatoes, gravy, peas, potato cakes, coffee and wine, all prepared, cooked and served by our own little radio group. We spent all afternoon at it, with plates, silverware, kitchen and kitchen utensils, and wine glasses courtesy of the local populace, though they didn't know we were usurping their housewares and more than making ourselves at home. Someone found a 50-gallon keg of wine and it didn't take long for that to disappear. I think most of it went that night—at least most of the section soused into an unnatural sleep.

It's surprising how trivial things can make human beings super happy. Among the bric-a-brac in the Hotel Haus Kipper we found an electric train, complete with signals, tunnels, a transformer that worked, and lots of track. We set it up in our room and there wasn't a better way to attract visitors, which of course, wasn't what we had in mind. We, too, were fascinated by the operations of this simple gadget. We had the track running all over the room, under tables, chairs and divan, and partly out on the balcony. We put the tunnels and station houses in appropriate locations and even erected an incline and a bridge. There were times when practically the whole complement of radio section personnel filled the room, men sitting on the tables, draped over the divan, sprawled on the floor, kibitzing about how to organize and run the rail yards. From time to time, each man came in individually to play with the new toy. As for me, well, it'll probably do me no good ever to get my kid an electric train—he won't see much of it before I wear it out. Anyway, we spent our time in Bad Kreuznach with the train, a very serious pastime. Were we kids? Of course. Unabashedly.

Enduring Lesson Number Nine: *When you want to go back to your childhood days, go. Don't give a damn what others think.*

This interlude in Bad Kreuznach helped to take our minds off how the war was progressing. The Americans were approaching Berlin, the Russians had been met, and now the Allies had begun to pierce Hitler's inner fortress, the stronghold of Berchtesgaden. When our push began, back in Merlenbach, we yelled in glee that the war would be over in a few days and we could all go back home. Now that there was still some war left, we began to settle into the new routine—occupying Deutschland.

ENDURING LESSON #9

"When you want to go back to your childhood days, go. Don't give a damn what others think."

One day, someone brought up a quaint little thought: What if we were relegated to the job of occupation here *after* the war? For thinking such a blasphemous thing he was thrown out on his ear. But I don't suppose I would have minded too much, for there were indications that occupation GIs might get a chance to attend some of the foreign universities like the Sorbonne, or perhaps even Oxford. This would have been an ideal scheme for obtaining a cultural intellect which I had heretofore not yet had time to attain. But time would tell. It usually did.

One of the most beautiful buildings in Kreuznach was the Kurhaus Palast, with its Schauberg Theatre, replete with furnishings on a grand scale. There was no better place to set up a Division Service Club, so the Trailblazers' Rendezvous came into being. We didn't pay much attention to this angle, however, inasmuch as our quarters were almost as elaborate. It was in the Schauberg Theatre that we saw our first movie since leaving the States. The Division Band played several excellent swing arrangements before the picture. The film, *Our Hearts Were Young and Gay*, was based on Cornelia Otis Skinner's travels in Europe in the 1920s. One part struck us with amusement—a scene that depicted views of travel posters reading, in a row: SEE BEAUTIFUL NAPLES… VISIT THE GORGEOUS RHINELAND… WELCOME TO JAPAN.

It was the last one that particularly shredded my soul. Next month I could be in Japan. We saw other movies and some good programs by the band while in Kreuznach and life was decidedly pleasant. There was one more little item that made life here less dangerous, at least in my case. The doorways of the Hotel were seven feet in height. This kept my steel helmet from getting any more dents in it that it already had. At that, my head

"You must learn not to come through those doors so fast!"

had received a number of Purple Hearts when I was foolish enough not to wear my helmet in French houses where we had been billeted. Even if you have a helmet on, it's a dizzying experience. I don't know if you have ever walked straight into a low lintel and smacked it with your steel helmet. Your head becomes the clapper in a bell and you reel back, staggering and disoriented for what seems like hours.

Most of the city of Bad Kreuznach reposed on the south bank of the Nahe River. A high cliff rose 400 feet from the banks of the river to the north. From this vantage point, an excellent view of the city could be obtained. We saw countless shingled roofs and tree-lined avenues. On one big building had been painted a huge red cross. Spring thunderstorms had come now to the Nahe basin and the trees had come out in full leaf. We could even see gardens of lilacs and tulips in full bloom.

Just below the edge of the cliff and stretching down the steep embankment to the river were vineyards. There was an old castle overlooking the town and we explored it. It had been lived in and apparently was so now, but the war had not done it any good and much was in ruin. In one cellar, we found big wine kegs, big enough, we estimated, to contain several thousand gallons. I suspect this could have been the city's water supply if the Nahe ever went dry.

We saw a lot of the plateau behind the cliff and had many occasions to look out upon the town below us, for this high vantage point also offered excellent radio transmission and reception conditions. We erected a CP tent and installed our radio control room. There were two radios, one NCS (Net Control Station) of the Regimental net, and the other connected with Special Troops. This meant that two of us had to be on duty at a time, so it was a social opportunity for plenty of gossip. Manning radio stations never got dull, for we could jabber to each other, and for the reason that we worked in six-hour shifts around the clock. Work times always changed. We had a little stove up there on which we brewed coffee, cocoa and tea, and even prepared C rations when we got too ravenous. We changed shifts in Peep No. 5, an undersized Jeep.

There were two routes by which we could get up on the plateau: one by the normal, less circuitous road, the other by a precipitous series of switchbacks that forced even the little Peep to back around. We took either one, depending on our mood or whether or not we were late. Telephone connected us by line down the embankment to Message Center and other points. The ack-ack boys set up some guns near our tent so we felt secure in the protection, though if they'd ever fired once without letting us know, we would probably have jumped clear off the cliff, tent, radios, C-rations and all.

After a while, we didn't have too much traffic on the air so we began to send practice messages, always identified by SVC at the first of the message. This meant dry-run and to put the message in the wastebasket. One, however, that should have hit the wastebasket, didn't. I got the message from one of the Battalions:

BT. GENERAL BARNETT ORDERS ALL UNITS OF 3RD BATTALION, 274TH INFANTRY TO PREPARE FOR DEPARTURE TO R________ AT Ø8ØØ THURSDAY MORNING. BT

Not seeing the SVC designation, I initiated the proper action and whisked the message off promptly to message center for action in the high command. That day all the men and officers of the 3rd Battalion, 274th Infantry packed up and prepared to move. The Battalion Commander waited and waited for the final confirmation and orders to move. They never came. So he phoned General Barnett to ask what was up and got red in the face when told that no such order had been given.

I went sizzling straight onto the carpet. I explained that the message had obviously been a practice one, and that either the sender did not send the *SVC* or I did not receive it. Everybody laughed it off, but it sure showed the power of radio in Uncle Sam's fighting forces. If we make a mistake the whole Division sinks or swims! (And for a while there, I thought I was sunk.)

Late in the evening of April 12th, Message Center called and said: "The President is dead." And that is how the two of us on duty in the Radio tent on the cliff got the news. It was a shock not many of us got over easily. The *Stars and Stripes* rushed out an extra edition, black-bordered and with three inch headlines to announce the news. There wasn't much laughter around when I got back to the Hotel after shift. Everyone had his own thoughts about the Commander-in-Chief. So did I: "Not only has a great man died, a great peace for all humanity might possibly be staggering on the threads of hope and fulfillment that he sacrificed his life to construct. Nobody can take his place now, but it is the world's fervent prayer that his successor will honor his cause with nobility and diplomatic intelligence."

On April 13th, we had our first parade in the ETO—to pay tribute to FDR. The Division Band played *Home on the Range*.

The war continued, and occupation kept on. From our vantage point on the hill and with binoculars we saw a Prisoner of War stockade some miles away in a valley east of Kreuznach. We were told that there were about 18,000 prisoners there and after looking, we didn't doubt it.

And it was here in Kreuznach that we noticed an article in *Stars and Stripes* about Ft. McClellan, Alabama, being quarantined for infantile paralysis. What a thought: we might have spent the war there.

I got a letter from my navy friend saying he'd made midshipman. At least he'd gotten somewhere after a number of us talked him out of the ridiculous notion of giving up V-12 and hitting the Great Lakes for boot training and then out to sea. He was still itching to get out on the water and it didn't help any when I told him how I was faring in France and Germany. I also mentioned that the war was going to be over before he got across, and that at least one of us was man enough to fight for his country. How we remained friends after that, I'll never know. But he knew I was hurting for him, so that helped.

Our food continued to receive the troops' approval. Breakfast, my favorite meal, consisted of hotcakes on one menu, and French toast on another, with non-appetizing powdered eggs thrown in. We had cereal daily, covered with fresh fruit when available. Sometimes the boys went out on hunting expeditions and we had venison. Sometimes there was nothing but Spam. This the cooks passed briefly over the flames a time or two and served nearly raw. Some of it reached the garbage can, but I confess I thrived on it, especially if I could find some way to singe it like bacon. In the hotel, the boys went right on drinking the wine they kept discovering in abundance. They drank it cautiously, though, after the story came out that several GIs had died and a couple were blinded by drinking poisoned wine in Nancy, France, a short while before.

We had time to catch up on our loafing, and I took time out from my Algebra course to do some reading. I had managed to get hold of a small book called *We Live in Alaska*, by Constance Helmericks, the story of a five-month canoe trip into the upper reaches of the Yukon. She and her husband had been students at the University of Arizona and had been considered fools for going on such a haphazard trip to the North country. But her descriptions were good enough to impress upon the reader just how pleasant it really was up there:

> *When we got Outside on furlough, people try so hard to show me a good time for what I've missed, staying here, and they feel so sorry for me. We show them pictures of our lives here, but they still can't understand why we should want to live here. They think it's just ice and snow and loneliness. But as for me, I couldn't live away from Alaska again. When I go Outside, that's when I'm lonely, among people I don't have anything in common with any more... Just look at it, where any place in the world today could you find a day more perfect than this? It was indeed glorious. We were pioneers. There was still pioneering to be done in the world, and we were glad. When you have once seen Interior Alaska leisurely from a small boat, and have stopped at every fish camp and every town, each with its different trader and teacher and missionary organization, the memory of this way of life will remain with you forever, and you will have seen and heard, tasted and smelled, Alaska.*

Such was the kind of reading that veiled my dreamy brain, half a world away from Alaska. If I were ever to travel in Alaska, and indeed I vowed some day I would, it would have to be in the far distant future. In the meantime, I was sightseeing in the other half of the world, and it was an education. The late President Roosevelt had once said: "I spent many of my earlier days bicycling around the Continent and gained from my experiences a vast fund of knowledge no classroom could have given me..."

Well, we weren't on bicycles but we appeared to be touring the Continent. Eventually the time came to say goodbye to the magnificent resort town of Bad Kreuznach, with its Nahe River, its Hotel Haus Kipper, its big cliff, its vineyards, lilacs and tulips. With a fond adieu, we headed deeper into the Gorgeous Rhineland.

THIRTY-TWO
FRANKFURT

While the 3rd Army headed into Austria and Czechoslavakia, we headed into the valley of the Rhine River and reveled in the scenery. Trees, grass, flowers, hedges, farms—greenery everywhere. And as far as rivers go, the Rhine may be classed as one of the most beautiful. It was wide, probably 500 yards as far as my inaccurate estimation runs. It was a deep blue, with tinges of emerald green along the edges. It was an important river, too, as could be seen by the vast number of ships that had been sunk in it and that now lay in ruin at abandoned wharves. It had been an ideal natural barrier for the Germans to hide behind, and of course, they had blown up all the bridges. We crossed on a pontoon bridge that to us was a miracle of US Army engineering.

We followed the Nahe River eastward from Kreuznach to the point where it runs into the Rhine, and crossed at the city of Bingen. On the other side we rode southward, upstream, along the banks of the river; it was early morning and the sun glinted off the trees and rocks and buildings as well as from the shimmering water of the river itself.

Past Winkel, Erbach, Eltville, and Niederwalluf to Wiesbaden. This last was inland from the Rhine, and a large city, a beautiful city, and a place indeed worth coming back and visiting some time. Then we arrived at one of the largest cities in southern Germany, a place we were to call home for a while: Frankfurt.

Frankfurt had been in the war. The air war. By the looks of the city the sky must have been darkened with B-17s and the bombs they dropped. When censorship permitted, I described it this way:

We rode into town and enjoyed the beauty of the suburbs. But when you get down town farther you feel as if you were in the Dakota Badlands. There are large areas of hills, composed entirely of bricks. Among these desolate hills are twisted girders of steel and miscellaneous rubbish. The street—bumpy with its numerous holes—has been laboriously cleared of the cascade of bricks. It's as if all the huge main buildings had just crumbled into nothing but bricks and dust. A couple of blocks in the distance can be observed the barren walls and skeletons of buildings that weren't so hard hit.

Then you ride through an area comprised mostly of hotels—averaging five stories in height and set closely together for blocks. As you ride along by these, it is amazing to see perhaps three or four in good condition, standing and occupied. But square between them are a couple of tons of piled bricks where one hotel, or whatever it had been, was destroyed completely. Yet the buildings around it are perfectly intact.That is a common sight. A row of neat houses and one between battered down. You might wonder how these long buildings had been singled out and destroyed. Perhaps the German Army had quarters in them and were shelled by artillery. Perhaps it was a stray bomb that missed the industrial district.

Who knows? Who cares? It still adds up to wholesale pulverization. I saw on one corner a beautiful entranceway—steps, foyer, a sign that advertised the "Kurhaus Hotel," and that was all. The rest of the hotel had been blown sky high. It was a once-proud city—with parks and avenues and a multitude of streetcars. That was Frankfurt, the wonderful city on the Main River, where now long lines wait among the rubble to get their daily bread—among half-bombed buildings and open, naked structural framework—and not far from a sea of destruction, the hills of brick. At Aschaffenburg, it was the same, and at Hanau and Wiesbaden. But at such resort towns as Bad Nauheim and Bad Orb it was different. The fast Allied drive had left these places untouched because the Krauts didn't linger to resist…

Our quarters in Frankfurt distressed us when we compared them with those at Bad Kreuznach. Here we had no private hotel but a military barracks that had only recently housed elements of the German Army. It was a big place, with ample room for our whole company to set up as a compact unit. The Radio Section selected a long building and we all took our places in the barren rooms. Joe and Mike and I took an end room with a western exposure opening out onto the street. There were double bunks but no mattresses, so we once again resigned oursselves to sleeping on boards. The radio control room was set up near the entrance and although pleasant enough, our roving team didn't see too much of it. The Division newspaper once ran a picture of this control room with the caption: *Here's the 70th Division radio control room from which is maintained communication to higher and subordinate units.* There was a short article alongside the picture that read:

570TH RADIO HITS PEAK

The Radio Section of the 570th Signal Co. has acclaimed itself as handling more traffic during the past two weeks than in any two-week period since starting operations in the European Theatre. During the two-week interval, the Section handled over 7,000 groups of code and clear text, which far exceeds the communication traffic before handled. Reason for the large amount of traffic given to them is the fact that divisional units and higher headquarters command posts have been located at long distances from Division CP. Telephone, telegraph and teletype communication facilities were not always available to the distant stations.

The Kowski-Morrow-Sutton team didn't flourish as much in the CP at Frankfurt as it did on field assignments with other Division units. One of the first things we did after we arrived in Frankfurt and selected our quarters was to go out to the Artillery and set up Wrangle's station (code name for the Artillery net) in the field at Bad Nauheim. We got there in mid-afternoon and were impressed by the grandeur of the scenery. Von Rundstedt had had his headquarters here at the time of D-Day, 1944. We drove into the city park and began to get our radio set in operation. We got the antenna up all right, but try as we might, there was no making the set turn on. What could be wrong? We worked for what seemed hours, we took off the back, side and top of the set—looking into everything except the battery box where, we learned later, the trouble existed.

"Then why don't you hook up your goddam antenna?"

Finally, with nightfall, we gave it up and went to the hotel where the Artillery CP had its radio on the secondary net. We radioed in and asked what in the devil to do. Eventually the answer came: come back home. Joe and Mike were about to drop dead from exhaustion so they curled up asleep on the long ride back while I drove. It was cold, and I huddled in my overcoat peering into the dim road that was lit by blackout lights. Finally, I began to realize that by now we must certainly be coming to Frankfurt and that somewhere here I should turn in or I might find myself in Paris the following morning. Pleasant as that might be, the thought of being lost was not. Buildings began to loom up in the darkness and presently an MP came out into the road and flagged me down.

"Say," he said, "did you pass a big civilian truck with semi?"

"Nope, don't recall having passed anything much between Bad Nauheim and here."

"Well," he continued, "one shot by here and I gave it the signal to stop. It didn't, so I fired 14 shots into 'er and still they didn't stop. Oh, well, we'll find it… Thanks."

He was off the truck before I could ask him where the 570th Signal Company was in this darkness. So I put the truck in gear and drove on down the road, on down one block—and pulled into the barracks area. It's embarrassing not to have any idea where you are and then find you're practically standing on the spot you were looking for all along.

When we got back to the company, we found out next day that we really should have stayed in Bad Nauheim, radio or not. For now, details were falling thick and fast, both guard and KP. One day when I was on guard, a little kid began to pester me and I gave him curt orders to get the hell away from there. He got mad.

"Tomorrow," he said, in German, "zee German planes, they will come, and alles will be kaput here." He went away.

I reflected that alles could not have been much more kaput than it was.

At this time, fraternizing with the German people was unlawful, which made it popular. Several of us once took a short walk along a lazy stream near the barracks and came upon GIs and frauleins swimming and making love all over the place. After seeing some of the frauleins, one could understand why the troops wanted to break the fraternization rules. At least it would be a pleasant way to get court-martialed. As a neutral observer, I vowed to have a closer look at these social interactions when time permitted.

A commendation finally came down to us from our old XXI Corps HQ. It said:

1. It is my desire to commend the 70th Infantry Division on the capture of Forbach, Styring Wendel and Saarbrucken and on the crossing of the Saar River. The successful completion of this mission is one in which the 70th Infantry Division may take justifiable pride.

2. The courage of your men is evident in their victory in the face of the enemy's stubborn resistance and employment of natural obstacles.

F. W. Milburn
Major General, U.S. Army
Commanding.

Our own commander, General Barnett, added that all members of the team could be justly proud of this official recognition of the combat efforts of the Division.

A day or two after the embarrassing Bad Nauheim experience, they entrusted to us the job of going out into the field again and setting up a temporary station, this time at an enchanting and luxurious little resort town called Bad Orb. Probably we spent more time than was actually necessary in selecting a suitable house for our operations but we had to have a high place for best reception and transmission.

We found a nice house on a hill—it had a charming view of the village, the forest and the valley, but the interior of the place wasn't exactly to taste. The next one we saw from the outside would be perfect, so we decided to take over.

That sounds simple, but we had problems. Times now were such that we couldn't just walk in and seize the place—there were formalities in kicking a family out. We had to go to the Burgomeister, in effect, the Mayor of the town, and the military commander of the town. Both of them got together and notified the family living at the selected site that they must move out by eight p.m.

As it turned out, the owner of the house had been an influential VIP, a Nazi BTO and an ardent Hitler fan. He went around all afternoon and evening trying to pull strings to keep us out. Apparently all this had failed, so he resorted to one last fight—he would bluff his way with us and play upon American sympathy.

We, however, were unaware of all this, and a few minutes before eight we took our truck up to the house, taking a road that led along an open field which separated the house from the fringe of a thick forest. The ground was somewhat wet after a recent rain and it took us some moments to get near the back yard of the proposed house, where we would set up our antenna. We got out to open the gate to the yard, and were met by the owner.

He attempted to feed us the same line he had given to the town bigwigs and told us, falsely, of course, that he had found out it was illegal for us to move in.

Mike listened politely to the man's outpouring of German and then sidled up to him and said softly but authoritatively: "Sie haben fünf minuten." [You have five minutes.]

This unexpected belligerence shocked the Heinie, who promptly decided not to play around with these American boys any more. He thrust out his hand in a gesture of friendship and said: "Ja, alles gut," but we were returning to our truck and preparing to move in. During the next few minutes, we moved in while the family moved out.

The home, modern in every respect, had large glass view windows framed by garden plants. The floor was thickly carpeted, the furniture as fine as marks could buy. We got our remote set-up hooked in and when everything had been taken care of and we were in communication, we started to prowl around.

For the German's sake, we shouldn't have done this. We found a number of things that it wasn't good for us to find. In the attic were several Nazi uniforms and many pictures of the son, a sergeant in the Cavalry. There was a stamp album in which the greatest pleasure had been derived in glorifying those stamps with Hitler's picture on them. These were innocent possessions, I suppose, and common to anyone who holds faith in what his country is doing, whether or not it's right, and we might not have been so badly affected if we had looked through the attic and quit.

We went down to the basement.

Here we found an empty box with a large red cross on it and, in English, these words: "To the International Red Cross Committee. Contents, 4 Prisoner of War Packages." Adding insult to injury, we found a large American coffee can, and an American joke book. Someone had diverted that PW package and its contents away from captured American soldiers and sent it to this house of an influential Nazi in Bad Orb. We discovered something else on the main floor of the house. In a cabinet by the door were two cases of German cigarettes.

While in France they will give food for cigarettes.

If you had never seen livid GIs, you would have seen them when these matters became known. Consequently, we had no scruples next day as we left. What happened was not our intention. But it happened. We had backed our truck downhill into the yard so that the radio transmitter could be close to the operations room. The ground was still wet. Add two and two and you get stuck. We ground around and slid from side to side not really caring where we went, so long as we could get out. But it was no use.

We called an ordnance truck to come up and get us out. The driver came in by the field out in back and approached the low brick wall that enclosed the back yard of the house. He stopped and got out of the truck to survey the situation, see how badly we were mired, and figure out how to get a chain on our bumper. He said he didn't want to damage their brick fence if he could help it.

I guess we should have kept our mouths shut. The enemy was gone and all but defeated. I readily acknowledge that it is wrong to punish innocent civilians loyal to their causes and leaders. But our minds were filled with the faces of close friends whom the Germans had brutally gunned down. Friends whom we would never see again. And in the back of our minds were the faces of those soldiers' wives and mothers crying and crying and crying... The result was almost inescapable, though perhaps wrong, wrong-headed, and unforgiveable. What happened was not the work of benign victors which we were supposed to be. It was a compulsive act of unnecessary revenge in a world where the psychological makeup of fighting men had been wrenched by the pains of war...

So, we just kind of let it slip what we'd found inside, and the driver listened carefully. He was a tough guy, not endowed with a lot of patience, and I feared for what he would do. Accordingly, when I saw the look in his eyes, I stepped back a few paces. Then a few paces more...

Without a word, he went back, climbed in the truck, slammed the door, raced the motor, and let out the clutch. There was a terrific crash as bricks and mortar flew all over the yard. Sod, grass, flowers, shrubs and trees were uprooted and thrown ten feet on either side. His tires dug deep holes in the gardens as our vehicle was slowly withdrawn.

And that is how we got out of the back yard of the Nazi house in Bad Orb. The house with the American prisoner-of-war packages in it...

It is not, however, to be assumed that all German people were as well off, or well-connected, or devoted to duty as the Nazi in Bad Orb. Indeed, quite the opposite, as we found one day during KP duty in Frankfurt. The time of morning came when we must take the garbage out to the nearest dump. We rode in an open 2½-ton truck and when we got there saw a number of shabbily dressed civilians at work. They were rummaging around in the smoldering trash heaps in a pitiful attempt to salvage something that shouldn't be wasted.

They bent over piles of cans and ashes and miscellaneous rubbish. Each had a burlap sack into which he or she would stuff some dirty object that might be edible or otherwise usable. At the sound of the approaching truck, they jerked up and looked our way. Seeing that we were on a kitchen truck with cans of garbage, they scrambled over piles of trash and ran toward us. They stood nervously while we backed into position.

We brought the cans to the rear of the truck, hoisted and overturned them, sending a cascade of sloppy refuse overboard. At once, the people—men, women, children—lunged for the garbage heap and began to sift.

One came up with a piece of dirty bacon, her face a picture of joy. Another found a coffee bag with a few grounds still inside, and uttered some gasps of delight. An old woman began to scrape some powdered eggs into a little bag. We had to yell several times for them to move so that we could empty the other cans.

This was intensely painful for us. We tried not to look, but there it was. We tried not to listen, but our ears caught every little whimper, every little whisper of delight. We tried not to think about it, but are you kidding?

We returned the empty cans to the front of the truck bed, closed up the tailgate and left. As the truck rumbled over the dump, we watched the little group disappear in the distance. They grew smaller and smaller as we watched them scrape out a few morsels from an old Spam can, pick up another piece of bacon, sort the last of the powdered eggs from the raw pancake dough…

We kept seeing the sight long after we had rounded a corner and shut the dump from view. We kept seeing the dirty wrinkled hands of the old women plunging into the filth, kept seeing their tattered, smoky coats, sprinkled with ashes. Worst of all, we could still hear the gentle little mutterings that might have been pigs at the slop trough.

We thought about it, thought about it long and hard and wished others could have seen it—wished the whole world, the people of America, everybody, could have been there on the dump truck that morning. We wished some of the leading diplomats in the world could have seen it, or better, experienced it, known what it was like to grovel for food on a garbage dump, to scrounge around for a scrap of bacon, a handful of bread crumbs or a soggy fragment of toast. We felt as if we *should* have seen it, that in those moments we came to know a lot more about war and what it meant and why it should never be. These were Germans. Maybe our GIs had gunned down some of *their* sons and grandsons. They were simple human beings made out of the same kind of flesh and bone and blood that we've got, and they were only representatives of a long line of French, Belgian, Dutch and other people who had gone before them. I thought of them as guinea pigs to prove to someone that war was not the right way out of anything.

THIRTY-THREE
VICTORY

The first week in May, 1945, was a big week, for us as well as the world. On the evening of the first we were sitting in one of our usual bull sessions talking about everything under the sun, when a sudden racket and yelling in the hall sounded like someone was tearing the place down. We dashed out to see what was going on.

BBC had just announced that Hitler was dead.

On the 2nd, the Army newspaper *Stars and Stripes* came out with four-inch headlines to confirm that pleasant little fact. That night we went to a Special Service show presenting a number of acts, the most outstanding of which was that of four Russians singing and playing instruments.

On Tuesday, the 8th of May, 1945, we picked up a *Stars and Stripes* extra which had a two-word headline that filled half the front page:

NAZIS QUIT

The story under it, in big black type, read:

> *Unconditional surrender of all German Forces was announced yesterday by the German radio at Flensburg. Grand Adm. Doenitz, successor to Hitler, ordered the surrender and the German High Command declared it effective, the German announcement said. There was no immediate announcement from the capitals of the Allied powers, but Associated Press and Reuter correspondents assigned to SHAEF stated unofficially that the Germans had surrendered unconditionally to the Western Allies and Russia at 0241 Monday (ETO time). High German officers formally surrendered the German forces at a meeting in the big red schoolhouse which is Gen. Eisenhower's headquarters, the AP and Reuter reported.*
>
> *Although there was no Allied announcement, the British Ministry of Information said that today (Tuesday) would be considered as V-E Day. Not waiting for formal confirmation of the peace news, New York and battered London, beflagged as never before, began celebrations. The surrender, said AP correspondent Edward Kennedy, was signed for Germany by Col. Gen. Gustaf Jodl, chief of staff of the Wehrmacht. Lt. Gen. Walter Bedell Smith, Gen. Eisenhower's chief of staff, signed for the Supreme Allied Command; Gen. Ivan Susloparoff signed for Russia and Gen. Francois Sevez signed for France...*
>
> *Kennedy said that Gen. Eisenhower was not present at the official ceremony, but that the Supreme Allied Commander afterward received Jodl and Gen. Hans George Friedberg. They were asked sternly by Eisenhower if they fully understood the surrender terms imposed on Germany and if they would be carried out by Germany. They answered yes.*

There was an article on How Doughs Reacted, especially within the 70th Infantry Divisions. One said: "I walked into the aid station and a Joe tells me the war was over. I thought the guy was kiddin'. Boy, that's what we've been waiting for a hell of a long time."

Another comment by one of the 70th boys was more along my lines. He said: "Some guy comes in and yells that the war was over. I couldn't believe it. If I can get a 21-day furlough, I won't mind going to the Pacific." Japan, here we come! There was a cartoon depicting two doughs in a bombed-out building. One had a field phone in one hand and he was laying his other hand on the arm of his buddy who was just about to toss a grenade. "Put the pin back, Joe… it's over."

For us in Frankfurt it was good news, too. Next day the *Stars and Stripes* came out with big headlines saying WHAT NOW? But we were too giddy to care about what happened now. Everyone celebrated, in one way or another. Otherwise, it was a nice sunny, warm day, much like any other spring day in Europe. Things were quiet in the city, with the people going about their work as usual. We listened to the radio a while, hearing how the people in the USA and Britain were celebrating. Then we sat back and began waiting impatiently to see the Army movie that had been prepared in secret and sealed in cans for release on V-E Day: *Two Down and One to Go.*

Wars, of course, cannot be wound down overnight. With the end of hostilities, speculation began among the boys. We had an orientation which ironed out a few things but left more doubts. Censorship was not immediately relaxed. We discussed the GI Bill, of which we might soon make use. It was said that at the present rate of discharge, it would be almost a year before any large scale demobilization would be in operation. We heard about the point system and found that discharges required a total of 85 points. That's all. Just 85 measly points to go home and be a free man. I had 26. Brother, it's going to be a l-o-o-ong war…

One thing Hitler could not change in Deutschland was the working of Mother Nature. Fred Larsen and I took a hike one afternoon for a good look at a typical European pastorale scene. We strolled down a road from the barracks and came to a stream that wound through a grassy terrain of hills and fields. The May sun felt warm and we lay on a grass-covered slope for a sun tan—a popular pastime here. Pretty frauleins ran back and forth, enhancing the scenery. Off in the distance, a village sat among a border of leafy green trees, its red roof tops and chimneys poking above the verdure. The frauleins and little kids and some GIs were swimming, and it looked as if it might be an excellent pastime, though the water looked a little cold. Besides, we had no swim trunks, yet I doubt if that would have made any difference to the populace. A small boat, rowed by three kids, came down the channel. It skimmed briskly along the surface of the water, around the bend, and under the shade of a big tree hanging over the water, then around another bend and out of sight behind a hill. White and yellow flowers lent a Technicolor dazzle to the scene. Overhead all afternoon flew all sorts of aircraft, singly and in formation, some frequently rolling over and swooping down in flight. This was one of the softer parts of life overseas.

At Frankfurt began the publication of a little company bulletin called *Short Circuit.* Since unit newspapers were banned, this had to be small enough to be called a bulletin. And it was. Largely it kept us posted on current baseball schedules (in which the Radio team was one of the top winners), some news, and a few other trivial items.

Mike Morrow started teaching a language class in German, which was pretty well attended, either for want of something to do, or for the information and thus a quicker way to captivate these bouncing frauleins. As long as you're going to fraternize, they said, you might as well know the language. I was in the class, too, of course, but not exactly for this reason; when I went out fraternizing I didn't expect to need German…

Even in Germany, it was a pleasant sight to watch some of the return to peacetime normalcy. In a particularly sentimental mood, I wrote: "I looked out the window the other night, the first night the blackout was lifted, and perceived myriad twinkling lights from windows flung open. Reminded me of *When the Lights Go On Again*… all over the world. It was like looking into the eyes of freedom and I knew that all over Europe the horrible blanket of war had been raised by the hands of Peace; and the free world can sing out again louder than the voices in the Night on Bald Mountain. For the Eternal Night is dead."

Eventually, we got to see the film *Two Down and One to Go*, meaning that the Germans and Italians had been defeated and we must now turn to the Japs. The picture explained things… all too clearly. I smiled and made the best of it, resigning myself to a trip to the other side of the world.

The Services announced plans to leave only some 400,000 troops in the occupation, thus shifting 1,300,000 to places unknown. I figured I might get a furlough home before going to

Japan. They also said that those who stayed would be able to take advantage of large scale operations in the way of athletics, recreation and education. There would be contests that stimulated theatre arts and writing; there were plans for teaching music, and other ways of being active.

So, either way, 'twouldn't matter. However, the sooner the discharge came, the better. We did have one scare. The *Stars and Stripes* came out with an article entitled "War Department Designates Essential Skills, Tough Luck for Discharge." The article listed those skills, including radio operator, and added: "In line with this, all personnel so classified, either skilled or unskilled, will not be released from the Army even if their adjusted rating scores are above the minimum 85 points..." It later came out, however, that MOS 740 was not so classified.

We got another scare when Lt. Sandifer upped our ratings to MOS 766 (high speed radio), but then we found that that, too, was non-essential. So we relaxed.

Thus the war's end found us at Frankfurt, the beautiful but battered city on the Main. We had found that occupation wasn't as dull as it had been cracked up to be, and we even began to look forward to whatever amount of it we had left. By the middle of May, we began to want a change of scenery, and we got it.

THIRTY-FOUR
DP

We went swimming in an exotic valley, where huge, stately trees came to the edge of a high-grass covered meadow, then the grass continued on down to the edge of a lazy stream that meandered through the meadow to unknown points west.

It was very good indeed to come out from under winter's cloak and, feeling secure from prying eyes, we stripped and had an excellent swim. When Joe and Mike became tired, they sprawled for a rest in the tall grass near the water's edge. I continued to flounder around in the cool, refreshing water, coming up on the bank from time to time to dive in again. It was during one of these trips to the bank that I stood there a moment before diving—talking to the boys. All at once Joe yelled that it might be a good idea to duck, because there was a pretty fraulein coming down the path and that she could see me.

I looked around, and sure enough, about 100 yards away, a pretty young lass walked toward us with no more apprehension about my lack of clothing than as if I'd been another cow in the field.

"Oh, so what if she sees me?" I said, a bit irritated, and the boys turned red for me. "Over here in Europe that type of thing that we in the USA might blush about is a common thing. Look at her, does she seem disturbed?"

She smiled. I turned around to face her and bid her a good afternoon. As she went along the path she came within twenty steps of where I stood. The boys peeked out petrified from the grass. I waved her a greeting and made some comment about it being a nice evening. She waved back, smiling, and returned that it was indeed pleasant weather. Then she went along her way. I would have invited her to tarry with us, but figured that Joe and Mike might die of fright.

This little episode took place on the banks of the Lahn River, which flows near the village of Dietz—to which the company moved from Frankfurt on the 18th of May, 1945. Actually, the first day, being confused as it usually was on moving day, we didn't go directly to Dietz but into a suburb that sits above the town—a castle and a group of houses known as Bad Oranienstein, presumably a castle once owned by William of Orange. I'm afraid I'm not sufficiently acquainted with my History of Western Civilization to recall at the moment a detailed story of this castle, but it must have seen its share of feudal excitement.

It was an impressive place. It formed a square, with the south side open, and this was the entranceway. But the place was big—the grassy square in the center over a block in all dimensions and the buildings that hemmed it in on three sides were three and four-story masterpieces of architecture. This was a typical thing on the Continent, it seemed. Since there was so little wood employed in the building of homes and business establishments, it allowed that stone and brick and masonry could be shaped into many picturesque and sometimes fantastic designs.

Besides that, this land had been occupied for millenniums, while the America we knew dated only from the 1600s, and as for me, living on the frontier as I did, in Arizona, buildings had only been in existence for a hundred years at most. Even the Pueblo buildings dated back less than a single millennium. Thus in Europe there had been many generations of masters from which to draw for extraordinary architecture.

The Castle of Orange, however, was not fully occupied, so we took over a part of it and set up radio operations. We began looking around for bunks. There weren't any. We found a piano in one of the lower rooms and at one point during our stay had a concert there. But much as we would have liked to stay in so glamorous a place, such was not to be the case, and early in the afternoon we pulled up stakes and moved a few blocks down the hill into the town of Dietz proper.

This, they said, would be a more permanent base of operations and the thought that we'd be here a long time made it even worse to explore our new building. It was terrible. The place appeared unclean and we figured the first thing that had to be done was a thorough washing from top (4th floor) to bottom. Maybe even a delousing. you never can tell about run-down places. A lot of trash had been left in the rooms, debris that we ourselves would soon be hauling out. We were told that the building had formerly housed Displaced Persons (DPs) that the Germans didn't much give a damn about and probably had thrown them some dog biscuits and water from time to time. At any rate, those DPs had gone, and we DPs moved in.

Joe and Mike and I began to grumble about the unpleasant aspect of living in this place when we were suddenly called out to go down to the Engineers and set up a radio station with them. Nothing could have been better news and I don't believe it took us more than two minutes to get all our possessions right back in the radio truck and take off, as fast as we possibly could, before they changed their minds. We hastened to get as far away from the DP CP as possible to evade the clean-up campaigns we foresaw.

That day we took off for Bad Schwalbach with the Engineers, and later moved to Bad Assmanshausen-am-Rhein, two of the most delightful stations in our respective Army careers… but here now, that's another story.

Shorty in the radio truck, Germany

We came back to DP after those forays, and found the experience not nearly as bad as we'd envisioned. We came back because it was unfair to keep the other teams around home base all the time so we had to take turns, and that meant coming back to Wyandot's horror once in a while. (Wyandot was the code name of the Company's home radio station at Division headquarters.) And yet, when you get right down to it, the place wasn't as bad as all that. The boys had cleaned it in our absence and had done a remarkable job. Our team selected a second story radio room and it became home. We had our usual bull sessions there and had fun talking about going home; we wrote letters and played cards and made life in the DP as pleasant as possible. But it was what went on outside the building, in Dietz and during our off hours, that was most interesting...

There was a swimming pool in a gymnasium building up on the hill. It was the most magnificent swimming pool I ever had the privilege of swimming in. The entire room was high-ceilinged, made of marble. The water was an exquisite emerald green, yet clear enough to permit observation all the way to the bottom in the deepest part. Many a pleasant hour was spent there in cool luxury.

Now that the war was over, much effort was made in giving the boys a chance to go sightseeing, or fishing or hunting. We went on a fishing-sightseeing trip one day, leaving shortly after dinner. We were loaded with fishing tackle and K rations. The roads were good all the way and we must have driven 40 or 50 miles before we came to a heavenly little stream flowing below the road in the heart of magnificent mountain country. We drove the truck up a side road, hopped out and chopped some poles. It didn't take long to string our lines and hop across the road and down to the stream. Eagerly we flipped our lines into the water and braced ourselves for the time when the first big one would hit. We could see a 20-25-inch fish come roaring out of the stream and land at our feet. In two hours of fishing we got a total of three nibbles.

A little kid directed us to a swimming pool nearby so we had a quick plunge to cool off from the afternoon heat. We devoured our rations while we were drying off, then dressed and drove on down the valley. After about five miles we perceived an old castle perched up on a hill, with a commanding view of the countryside. Nothing would do but that we go up and look it over. There was a small village nearby and we finally learned that we must find the key to the door of the castle since it was now more of a prehistoric ruin. Apparently the house nearest the castle contained its caretaker so we went to him and obtained the key, that is to say, coaxed him into going in with us as a kind of conducted tour. While they were getting ready, I went in the house for a drink of water and was almost knocked over by a couple of dogs prancing around locked together, victims of the love season in German dogland. The woman of the house was so embarrassed, she turned red as a radish. I suspect the dogs got a thorough flogging after we'd gone to the castle.

We circulated all through the castle, except in the lower portions where a labyrinth of dungeons could only be penetrated by those in the know of the place. We climbed rickety ladders into lofty towers and looked out through portholes that had been used to fire on attacking enemies. It differed little from Indian cliff dwellings in the Southwest USA, for they, too, had portholes through which they could pick off an approaching enemy. There was another similarity in that this castle had been built in 1190, and that was about the time cliff castles were being built in the US West, too, long before Christopher Columbus was even contemplated.

We were in Dietz on D-Day plus 365, which was declared a military holiday. I celebrated it by going swimming, going to sleep, and being Assistant Sergeant of the Guard. The days in DP were marked with such details as being on the radio, the switchboard and the telegraph net, and these weren't bad times, though without the excitement of battle, the days sometimes passed slowly.

One day, we got our biggest load of PX supplies and had to stand in line two hours to make purchases. We each got an allotment of 14 candy bars, 12 packages of cigarettes, a can of peanuts, a can of grapefruit juice, wafers, gum, toothpaste, soap, towels, film, cigars and other little odds and ends. It was nice to fill up our personal stock again. We had some personalized stationery printed up in a town called Bad Ems, located about 25 miles away. While we were in Dietz, our company newspaper was printed in Bad Ems, too. It dealt largely with intra-company, intra-divisional sports and all the activities that kept the boys out of mischief. There was a picture contest in which they selected THE GIRL WE WOULD MOST LIKE TO OPEN COMMUNICATIONS WITH. The I & E Section (Information and Education) started electricity classes, which, by popular demand, were selected as a starter for an increasing educational program. There were also plans to produce a stage show based somewhat on thc lines of the Folies Bergères in Paris. Ours was titled "Folies Brassieres."

Even with all these activities, the boys did not keep out of mischief. Though still strictly against the law, fraternization became one of the most popular pastimes. As a matter of fact, it became so popular that the supply section nonchalantly issued a disease-preventative kit to everyone in the company. Just a safeguard. However, most of the boys had been away from women so long and had been so busy fighting a war that now the pent-up I-want-some-lovin' feeling had to be expressed. And it was.

To find out how much, I decided one evening that I'd walk down Fraternization Lane to see what went on. This was a particularly beautiful and heavily forested area that stretched along the banks of the Lahn River. It was a place where town girls and GIs met in secluded rendezvous. A lovely place, precisely the locale that could be so inducive to the art of lovemaking. Besides that, it was Spring, and you know to what a young man's fancy turns during this critical period. I walked down with a chap named Al Simmons—a bull in search of heifers if I ever knew one. He had a message to take to some fraulein, just the sort of situation, he said, that could serve as an introduction for us both.

Her name was Maria. So we walked along the banks of the river in the late afternoon sun, inquiring here and there as to her whereabouts. We saw a Jeep load of MPs coming down the road, which was merely two weed-grown tracks alongside the stream, and we tried to look as innocent as we probably thought we were, but I don't think it looked becoming. We found out, however, that the MPs patrolled the area with tongue in cheek when it came to mixing with the frauleins. Undoubtedly they were here themselves during their off-duty hours. They couldn't blame us. We couldn't blame them. Non-fraternization had its principles, to be sure, but even though the troops knew they were breaking a sensible law, they loved breaking it just the same.

As the sun lowered toward the heavy forest to the west, we finally came upon Maria. She had a friend with her. Al gave her the message: Anthony couldn't meet her here tonight because he was in jail for speeding in his Jeep this afternoon. And somehow, we just couldn't stand for the girls to be so lonely during the evening, so we sat down to talk with them. Al took Maria. I got her friend, a better choice, to my way of thinking. We introduced ourselves and my girl said her name was Taya (the way I choose to spell it), which was a contraction of Dorotea. There had apparently been some French lineage with her, and I figured she knew something about making love. At least we could swap notes on German and American techniques...

I have to confess that it was nice with Taya there in the cool breezes of late afternoon. Tall grass fronds lifted their stalks in silhouette against the deep blue sky. No one could see us. We were at least filtered from the eyes of the MPs. The Lahn rambled lazily along, rippling here and there. Neither of the girls knew French or English, so we had to get along as best we could in German. And we got along all right.

Soon we quit the community talking, distanced ourselves a bit, and devoted our attention to the respective girls.

Taya worked her way into my arms, enjoying the pleasure, I assumed, of these *soldaten Amerikanischen*. We talked. She told me she was a nurse, had been in the war, and was now serving at a small hospital in Dietz. Admirable occupation, I told her. I tried to tell her I was from Arizona, but she knew as much about Arizona as I knew about South Africa, so that didn't get us anywhere.

I looked at her, from an objective point of view. She was well-formed. Maybe a little on the chunky side, but well-formed. Stacked, as the Americans would say. She was brunette, with little brown eyes that had a way of looking at you and saying, "Hello, big boy!" She looked up at me now and then, and her eyebrow gave a high, twisting arc and snapped a salute in my direction. As she talked, her eyelashes flipped in delicate arcs that seemed like palm fronds waving in a dreamy, tropical breeze. The lines of her graceful form seemed to slither majestically along among the stalks of high grass that surrounded us.

She placed her shoulder against me and nudged, just ever so gently, so peacefully, so much with grace. Her little nose was poised in delicate exquisiteness above beautifully curved red lips that moved with ecstatic loveliness when she spoke. The afternoon became exotic and I began to say things she couldn't understand.

Faintly in the distance, I could hear noises that seemed to say that Al, the bull, had found his heifer.

"It's good to have lots of outdoor recreation for the troops, general."

Taya took my hand and caressed it. Then I began saying things *I* didn't understand. She had a magnificent bracelet which she assisted me in examining. Then she put my arm around her and moved in closer... And with a deft movement that spoke of experience, she gently lifted my hand and dropped it on her ample breast.

Now this is the point where the movies dissolve away to the scene next day, but I must tell you exactly what happened. And there is no point in drooling as though to receive a pulp fiction episode from *French Romances.*

I realize that in the soft grass, on a lazy afternoon, beside a murmuring river, it is logical to assume that the natural progression of events took place. But when Taya placed my hand on her breast, a voice in the back of my mind became very powerful.

How many times has she laid other hands on that breast? How many times has she put other arms around her? And what germs does she harbor that maybe even she doesn't know about? Not even a nurse...

The graphic training films showing lesions and blisters on men who had gone too far with the wrong persons came back to haunt me. I said to myself that if I were ever to get married, I should start respecting my future wife right here, and right now. If I were careless, I could transfer those germs to her, and to any children she might have.

This may sound insufferably prudent, but to me these were more the facts of life than any passing conquest on the way to sexual fulfillment. I suddenly decided that I would withdraw from this situation before it went any further, that I would not yield to the passions with which so many human beings have been naturally overendowed.

I looked at my watch, gasped in surprise, crossed my face with a frown and... Well, I don't need to go into any further details, nor can I calculate whether Taya was disgusted, disappointed or dismayed at my early departure. She had been the aggressor, so to speak, and I tried to be a gentleman about the whole matter, thanking her for a wonderful interlude, bidding her goodbye, *auf Wiedersehen*, and *bonne chance.*

I walked back alone. Al already had a wife. I just hope he didn't transfer anything to her from Maria. I had discovered exactly what I wanted to know: the events that took place down Fraternization Lane, on the graceful banks of the Lahn River... in a wartorn world of loneliness...

THIRTY-FIVE
SCHWALBACH

It has already been brought to your attention that time spent in the CP DP at Dietz wasn't always as pleasant as it might have been, even in our own room 90.

It was thus with pleasure that we were sent out to subordinate units to set up radio operations. The most opportune time to go out to such a place was when we first arrived in Dietz and we got out before the clean-up detail started throughout the building. Such was our fortune when we were assigned to a place called Bad Schwalbach.

Away we went in our radio truck, though magnificent forests remindful of Arizona. I tried for a while to imagine myself back among the ponderosa pines of Coconino County, and I almost succeeded when we passed a railroad bridge. That is, where the bridge was, the river is.

I had never seen so many hotels in all my life, not even in France. You wondered if anyone ever stayed home. We really got a nice place in Schwalbach, in the Hotel Herzog von Nassau. The Engineers welcomed us with open arms, and there were a number of the boys whom we knew quite well. After chatting for a while, they showed us our radio operations room, a delicately furnished 18-Century chamber on the first floor, and then to our quarters, on the fourth floor. We had separate rooms, with garret windows opening out onto a view of the city. I figured this was where I was going to get my exercise—climbing those flights of stairs to the fourth floor.

All the furnishings were courtesy of the owners of the hotel, who kept their distance and pretty much out of sight in a room in a wing of the building. We didn't see much of them, nor did we need to. We got along very well. The radio room was installed with full technical equipment and remote units.

For a chair in which to sit during duty hours we had a high-backed, red plush, velvet affair that was highly inducive to sleep between midnight and dawn. For the most part, however, we managed to stay awake, if only for the luxury of enjoying the surroundings while we had them. And enjoy them we did.

Meals with the Engineer Company in Bad Schwalbach were extremely good. They were served in the secluded, heavily landscaped patio or courtyard on the side of the hotel opposite the main street. We sat among the trees, shrubs and garden flowers for our meals. There were statues on fountains in the garden, most of them, of course, nude. One of the boys wanted my picture on one of the pedestals, leaning against the statue; I had to watch where I put my hand. There was a 50-gallon canvas bag with spigots on it that served as our source of drinking water. I was leaning against it one day mediating when some voices came to me.

"Where can I find a drink of water?"

"There's a big one over there."

"Which one?"

After that, I sneaked in to get a drink of water when no one was looking.

Pleasant days passed at Schwalbach, more like a vacation than a war. Up in my private room I worked on my algebra course and devoured sweets from packages that had come in the mail. There was also time for reading, even a little on duty at the set, since not much traffic was being handled.

I walked around town one day, observing that the most popular business here, as well as in other German towns we'd passed through, was *Damen Und Herren Friseur*, or hairdresser. There were a number of drugstores, or *Apothekes.* There didn't seem to be much activity.

Oddly enough I found a photograph shop that had been approved for all Allied personnel. I wanted to have a professional picture taken and walked up to the door. The sign said: *Dienstag und Freitag, Nachmittag, geschlossen.* Obviously the place was closed on Mondays, too, so I had to wait until the next day. I got the pictures taken, asking the photographer, in French, to have them done *toute suite*, in case I had to leave hurriedly. He said he would, but that since there were so many soldiers here he had *beaucoup travail.* I got six small pictures for eight marks (80¢) and three large ones for $1.80.

Walking over to the *badhaus*, I looked in on the mineral baths. There was no activity on this particular day, but the boys said that mineral baths they'd had were particularly stimulating. Each separate bathroom had a metal tub, adorned with all sorts of regulating and heating devices. Some day, when they had hot water, I told myself, I'd take a mineral bath. I never got around to it.

On the 25th of May, censorship ceased and that old byline "Somewhere in Germany" ceased. Then we wrote freely about where we were and what we were doing. By this time, we'd been assigned to the Ninth Army, and had no idea what the Ninth Army was going to do in the scheme of things. There was the horrible thought that the 70th would be transferred to the 15th Army, whose duty was solely that of occupation. Our free combat rations of cigarettes and chocolate bars ceased, but this didn't hurt, since cigarettes sold at 5¢ a pack. We received notice that we could add two campaign stars to our ETO ribbon: Central Europe and Northern France.

The closest I ever came to getting shot in all the ETO happened right there in Schwalbach. And, ironically enough, it was my own gun.

We always carried our M-3 submachine guns with us wherever we went, and it came to be such a habit as to be comparable to wearing our helmet everywhere. One particularly nice day, a group of us loaded up in a "cattle car" and went down to a nearby theatre and saw a 16mm print of the Errol Flynn war film, *Objective: Burma!* It turned out to be one of the finest war pictures to come out of the conflict.

We hauled back on the truck when it was over and rode back to the hotel. A group of us strolled into the radio room to gossip a bit and perhaps discuss the picture. We laid our coats, helmets and guns down in a pile on the floor, which is where we normally placed our gear when we didn't want to be burdened with it. I went over then and sat on the window sill, that being the most convenient place to sit in the room at the time.

I didn't happen to notice that having taken this position it placed me in such a way that my gun was pointing toward me. I didn't even give it a second thought—by now we were so used to the weapons that we didn't ever bother to see where they were pointed. Of course, the gun was loaded, useless to carry it if it were empty. But the safety catch was on, as a precautionary measure.

As we sat around talking and smoking we were unaware that a near tragedy was about to strike. We talked and laughed as if everything were fine. Suddenly the room was filled with the loud report of a gun going off. The terrific crash welled up and reverberated around the room as though lightning had struck the antenna and the radio had exploded.

It took us some time to get our wits back after the noise had died down. Someone yelled: "What in the hell was that?"

We looked around. There wasn't an enemy in sight. Everyone in the room was apparently still alive. It took a little while to discover what had happened.

My gun had discharged. One shell. The slug had ripped into the wall beneath the window in which I was sitting. That wasn't all. The slug had ripped into the wall directly between my legs. When I made this charming discovery, everything went black.

"Farewell, fellas, it's been... swell... Oh-h-h, I'm going fast; everything is so dark... Give my regards to the boys back at the base... Tell them... tell them I... I..."

One of the boys said: "Oh, quit it, you silly bastard. You ain't hit."

So I came back from the grave, as it were, and examined the bullet hole... but not before carefully picking up the gun and placing the lock on it. Evidently that swinging up and down the side of the truck on the way to and from the movie had somehow flipped the catch off the safety position. The slug had bit deeply into the cement and plaster of the wall. And to think—that could have gone off when I had my leg a few inches to the left. Worse yet, I could have sat on the floor rather than in the window. Or this could have happened when the gun was flipping all about me in the truck.

On the other hand, it just couldn't have, shouldn't have happened at all. But it did, and I know of no better way to instill in a person the advantages to be derived from careful handling of a gun. Unless, of course, to get hit with it. The government came awful damn close to losing $10,000 insurance.

One afternoon a group of us went to the outskirts of Bad Schwalbach for an hour or two of swimming. It was an excellent pool, located in a small valley with a forest on one side and fields stretching up the other. The Reconnaissance section of the Engineers had just cleaned it out and made it fit for activity.

There were a couple diving boards, one up to ten feet. So we got into what swim trunks we had been able to scrounge and set out for the afternoon. We did not know, of course, as we advanced toward the tempting water, that it was fed by icy underground springs. But then, I told myself, what if the water is cold? I was used to this—had I not dived into Oak Creek back home once during the month of February? This pool couldn't be as cold as that, and would therefore be snap potatoes. Watch me show off how brave I am to the rest of the sissies here!

I flipped gaily up to the edge of the pool and without a moment's hesitation dived in. I dived right back out in a geyser of spray that would have put Old Faithful to shame. When I landed back on the bank, someone yelled for the boys to bring up an ice-pick.

Eventually I turned from blue back to a healthy flesh color and from then on treated the water with more respect. Oak Creek had been lukewarm compared to this ice pond. At least we got the heat of the afternoon sun out of our bones.

We didn't go through all of Europe without having ice cream, a popular commodity very difficult to come by. Somehow the cooks managed to get some made, and I had my first spoonful on the 28th of May, the first since way back in Myles Standish. About a spoonful was all there was, but it sure hit the spot.

Randall Foley and I decided that morning to drop into the casino across the street from our hotel and see what it was like. Surprise! We found a court-martial in progress. It was open to the public, but we were the only spectators. We sat and listened, becoming very interested indeed. It was a rape case.

The accused was a GI from one of the Infantry companies and the accuser was a local fraulein. The jury was composed of six Army majors and there was a prosecution and a defense, a recorder and an interpreter. During the hour we stayed there, four witnesses were called—three middle-aged frauleins and the accused's company commander. The frauleins all served to bear out the testimony of the accuser, insofar as they could without having been actual eyewitnesses. Then the victim was called to the stand and gave her story, one difficult to piece together since it had to go through an interpreter.

She described the GI's entrance into the house, which, they found out, was acceptable to her. He had apparently entered on the pretense of wanting wine. She had treated him well (and his partner, too, for he wasn't alone). Then the accused asked where the bedroom was and asked the woman to show him to it. At this point Randy and I leaned forward in our seats, clinging to every word.

The fraulein said that the GI had gotten her in the room and attempted to lock the door. By this time, she testified, she had begun to suspect what was going to happen and she became panicky. She got on one side of the bed to use it as a barrier between herself and the man. He didn't bother to waste time chasing her. He began to make certain motions to indicate what he had in mind.

"What kind of motions?" the judge asked. "Will you describe them please?"

She described them, through the interpreter and the picture still remained hazy—partly because the woman was embarrassed and halting in her remarks.

"Now," put in the major, "will you show us exactly what he did. Will you illustrate the motions this man made toward you that day?"

She got up and went through the motions. They appeared real enough to incriminate the defendant.

A recess was called at this point and Randy and I figured we'd better not spend any more time here because we were due back at the company. Just before we got up, one of the majors came down the aisle on his way to another room. He saw us sitting there and, as he came by, leaned over and said something to us. I remember his remark well.

"Juicy son-of-a-bitch, isn't it?"

Next day we heard that the man was found guilty and sentenced to "hang by the neck until dead," which at the time seemed a rather harsh sentence for the offense in question, and may have been a rumor. We never knew. And, of course, we hadn't heard the trial in its entirety. On reflection, however, I began to think of the reported sentence as rather logical. We were all supposed to be on good behavior, avoiding the atrocities that Axis troops had committed. If fraternization itself was a serious offense, what kind of a crime would rape be? We were quick to defend American soldiers against Germans. Hadn't they killed so many of our friends? But in this case, hanging would have been about right, sending a powerful message that Americans did not stand for such brutal behavior.

THIRTY SIX
THE RHINE

In Germany, the Rhine is an extremely handsome and picturesque subject for sightseers and photographers. There is color, both within the scope of the river and distinctly along its banks and the adjacent lands. It courses along smoothly at first glance, yet turbulent from within. Here and there, eddies and whirlpools and savage, slithering currents flow steadily and powerfully northward. Deep water is a rich blue-green, consistent and even-shaded through most of the width of the river, the only exception being along the banks, where the hues of emerald green predominate.

By all measures, the Rhine is a big river. Its flow must be well over half that of the Mississippi. Its width might easily have measured up to a quarter of a mile, and actually varied that distance very little. Its major tributaries in Germany are the Sieg, Ahr, Wiedbach, Mosel, Lahn, Nahe and Main on the upstream course to Cologne. The river has a small number of bars and islands, as many occupied with buildings as would permit foundations to be built. Along the entire course of the river, it seems, there are villages and castles and vineyards, quaint and jeweled in their settings.

From Cologne to Bonn, southward, the River meanders slightly on flat or gently rolling plains. It flows through rising foothills that eventually close in and form a canyon along the river. The high cliffs open now and then for a tributary that has, in times past, built up a delta deposit upon which cities have been built, like Koblenz. There is very little flat country from there on down to Mainz, but stop! Not so far down. We want to stop just around the bend before getting to Mainz and Bingen and investigate the little village on the east bank, located obliquely from the junction of the Nahe with the Rhine—an agreeable little village nestled on the river bank and squeezed in at the base of a tall, heavily forested cliff.

Here, the map will show you, is located the handsome health resort of Bad Assmannshausen-am-Rhein, or Assmannshausen for short; well, half short anyway. We had to be careful just how short we did make the nickname. And this was our next home in Germany. It was the first of June, 1945.

"*Assmannshausen,*" the folder says, "*Bekannt durch seinen versuglichen Rotwein und seine vielen Gaststatten. Von hier Zugang zum Jagd Schloss und Niederwalddenkmal. Dampferstation.*" [Assmannshausen is known for its quality red wine and its many inns and taverns. From here is access to the castle used to hunt, and the Niederwald Memorial. Dock for steamboats.] We liked it. We packed up our radio truck with all the equipment in Bad Schwalbach, did loop-the-loops on the roof of Hotel Herzog von Nassau to extricate the antenna moorings and take down the flat-top, then climbed in and shoved off, maintaining communications along the way. At 9:00 a.m. we drove out of town, motored along rustic valleys and villages, through scenic vales and beneath forested hills. We came to the Rhine and drove downstream a matter of ten miles to Assmannshausen. It was all I could do to keep my eyes on the road, what for the scenery, the war damage, and all the sights.

We arrived at 10:30. Our quarters were assigned in the Kurhotel, an attractively landscaped place with gardens in front and the Badhaus in back, located about a half mile down from the main part of the village, and actually on the outskirts. We were sandwiched in between the Rhine and the highway on the one side, and the railway up an embankment on the other. The hotel was exquisitely furnished, and the furnishings were still there. Our radio room and sleeping rooms were to be upstairs on the third floor, so we began to set up. For two hours we labored with the antenna, hooking one end on the farthest corner of the roof and the other end on a tall tree a short distance away from the building. We slipped and slid on the smooth shingles of the roof, and Joe slipped and skinned himself nicely while scaling the tree. The antenna must be exceptionally well installed here, for radio signals had to rise out of the valley up over the steep walls to get anywhere.

Soon we had the 150 feet of wire up and were back in communications. Then we got down to the business of seeing about our accommodations. Our room was just across the hall from the Operations room. In it were four lovely beds with clean sheets and pillows (kept that way by maid service while we stayed there), reading lights, tables, desks and a wardrobe closet. The view from the window, looking downstream, approached the sublime. Beneath the window was the sunken hulk of a cargo ship, with countless bullet holes in the stack and throughout the hull. Along the east wall were what seemed like millions of terraces of vineyards, each banked up with a high rock wall, a building job that must have required centuries of hard work.

On the west bank, across the river, grew a very dense forest. About a quarter of a mile downstream, a little cove ran inland a short distance, and on a promontory jutting out over the river stood an old feudal castle—Falkenburg.

The highway wound along the river's edge around the bend and out of sight on the east bank. No matter how much we looked or explored, we could not get over how the steep slope of the east wall of the valley had been utilized for grape-growing. Terraces of vines crossed the slope and were connected by a laced network of steps. It would have taken an hour to climb to the topmost terrace. From time to time, we saw farmers along the hillside, on different terraces, tending their crops. After seeing all this, we figured why wine was the national drink. Well, why it was *one* of the national drinks.

By the way, I've not yet mentioned the drink which became quite famous to the GIs who visited this charming country. This one was Schnapps, bottled dynamite that could lay you low in a matter of seconds. It sold for a high price, a carton of cigarettes bringing a small bottle. But a small bottle could last the drinker a long time, for he could get moderately drunk on a few sips: lightly happy in the same way that might ordinarily require a couple strong glasses of beer. Schnapps was only for the most hardy and for those who wanted to get it over with in a hurry. Gulp down a whole bottle and you would wake up next morning.

I got better on my German and managed to talk the hotel manager out of some souvenirs, one of them a large 7 ft x 1 ft map of the Rhine. I also got some pictures from him that would have done credit to any of the most enterprising tourists. He had a nice hotel there and was proud of it.

The lobby was well adorned with statues and chandeliers, a restful place at any time of day. Outside, a shell had knocked off a portion of the steps and sun room, but the grounds had been kept in good shape. Along the river within view were several boat landings, ferries in all probability, about a quarter of a mile apart. Most of these had been blown up. Half sunken, they were being eaten away by river currents.

Perhaps the most impressive sight from the Kurhotel at Assmannshausen was directly across the river where a castle stood precariously on a rocky promontory half way up the cliff, poking out of the dense foliage. This was the Burg Rheinstein, about which I knew little or nothing until the hotel manager endowed me with this description (given here with translation errors included):

Burg Rheinstein. On a precipitous rock, in a hight of 90 yards vertically above the Rhine-river, juts out of a solitary rocky vale the Burg Rheinstein, which is a gem of the German past with regard to the exterior aspect as also to the interior establishment. It seems to represent a tale of olden

times with all it's ivy-clad court-yards, open galeries, with it's gardens and the precious furniture. Rheinstein was called the Burg of the bailiffs and it served the archbishops and the electors of Mayence for a stronghold for collecting of customs. It is for the first time authentically mentioned in 1323.

Prince Friedrich of Prussia acquired the ruin and reconstructed it between 1825 and 1828. The Rheinstein is the first German Rhine-burg raised anew. Prince Friedrich was a sister-son of the Queen Luise of Prussia; he bequeathed the Burg to his son, the Prince George of Prussia, who left it to his nephew, the Prince Heinrich. It is still the private property of Prince Heinrich's family. The Burg is completely established with pieces of old furniture and valuable paintings. It holds a number of rarities and a rich collection of weapons and suits of armour.

On the rocky crag, with a lofty church huddled close to the main castle, it was an impressive sight. There didn't seem to be any manner or means of approach to the place, but the trees grew so profusely on the cliff that they might have concealed approach roads. We wanted to see it at close hand, but figured it wasn't worth trying to swim the Rhine.

We didn't stay long in Assmannshausen, but it was an old soldier's rest period while we were there. We walked around, hiked and explored, and lounged in the luxury of the room that was kept neat as a pin. I spent a little time pondering over something that had happened a couple of days before we left Bad Schwalbach. There had been a religious event and parade that passed directly below my window up in the hotel. I had jotted down a description of the event as it occurred, in the hope that some day someone might be able to clear up what it was. Here's the description, now give me the answer. What was up?

0830, May 31, weather partly cloudy. The parade lines up in this order: it is led by four young boys dressed in white capes and green skirts. There are two lines, first children, then women, then a few elderly men. Part way back are two flags, one white cross on green, the other yellow on blue. Half way back is a large banner portraying Christ on a yellow field with purple trimming. Farther back are blue banners and red banners, elaborately decorated, bearing religious pictures. The two boys leading each line carry banners. Then there are two boys in the middle, one on the left bearing a silver crucifix, and the one on the right carrying a prayer book.

The parade started at 0840 under the direction of a man in plain dress clothes. Small children walk down the center between the two outer lines, and they bear smaller banners. Almost everyone carries a prayer book and reads it while walking. Bishop and nuns walk slowly along in the center aisle. There are small girls dressed in white, each bearing a gift or flowers or the like. There is an eight-piece religious band. Children, girls in white and boys in red and white capes and red skirts, carry candles. Two red and white flags now, followed by four men who carry the corners of a canopy over some sort of a bishop or cardinal. The parade is ended by a double file of 18 nuns, then civilian women and men in the last.

The procession emerges from the church and moves down the street to gather in the park. Bells are ringing and there are large yellow and white as well as red and white flags flying from many buildings, including the church. There is much religious incantation during the procession.

And that's it. If that isn't enough description, then I cordially invite you to travel to Bad Schwalbach on the 31st of May next and see for yourself. It was an impressive ceremony.

On June 3rd came that inevitable, dreaded order to pack up and return to the CP at Dietz and our replacements arrived, glad to be out in the open air of Germany and away from the rigors of garrison life at Signal Company headquarters. I suppose it was turn-about-fair-play, and let some of the other teams have a chance to enjoy rural life, but it didn't make us any less disappointed to have to leave Assmannshausen. On our way back, we took a long time and made a number of detours so as to enjoy our last breaths of freedom.

We drove leisurely down the river. We thought of going back and dumping our replacements in the Rhine, but since this is frowned upon in our society we gave it up and tried to forget about Assmannshausen in the comfort of the Eagle Red Cross Club at Wiesbaden.

We stopped for refreshments and found the Club located in a massive, modern building with glittering decorations. There were trees out on the vast terrace and tables at which we might sit to snack in the afternoon breezes.

We did just that. There was the fragrant aroma of baking doughnuts, on which we thrived, and besides, some very attractive Red Cross volunteers—real, true-to-life American girls who spoke English. There were also reading, writing and music rooms, and facilities for sports like swimming, and for cultural enjoyments like concerts. I thought of staying here a couple of months, but the boys would only carry me to the truck, kicking and screaming, and thus save me from being AWOL. Oh well, I'd be going home to the good old USA in 1960, so all I had to do was wait.

We finished the drive to Dietz through handsome forests and farms of the Rhineland. I don't know what I was thinking of when I rounded a corner and came face to face with a huge civilian truck, but I do know what I was thinking afterwards, as Angel's wings flapped about my head. We missed the truck, but it was so close that where there was smooth Army green on the left side, now there was a handsome crinkle finish to match the radio sets inside. Color didn't return to our faces for about three miles.

Back in the DP CP I settled down once again to hard boards and a sleeping bag, to algebra, letter writing, swimming, electricity classes and fraternizing by the Lahn. I never saw Taya, and I doubted that she would ever want to see me again. Fortunately, none of us had even the remotest idea of what was about to happen shortly, so we had nothing to anticipate with growing anxiety. As far as we were concerned, the 9th Army would be reassigned on the 14th June, and that left us with the thought that we might go home on furlough, preparatory to being sent into the Pacific Theatre of Operations. As usual, nothing was definite, so we waited as best we could and amused ourselves with that which was available.

One of the things most easily available was the surrounding scenery, with which we could amuse ourselves. The company organized sight-seeing trips, largely confined to Sunday, and planned to make a series of trips to such outstanding places of interest as Cologne, the Alps, and perhaps over toward Berlin. A few of us looked forward to these trips in order to gain an insight to exactly what had happened to Germany during the war—how she had been torn to shreds in a war Hitler had started and hadn't been able to finish. In the first trip, we found exactly what we were looking for. This was Cologne. It was the 10th of June.

THIRTY-SEVEN
COLOGNE

There were ten of us. We climbed aboard a $\frac{3}{4}$-ton truck, rolled the canvas top back and started out in the cool of the morning for the 80-mile trip. It took us five hours to go, stopping at points of interest along the way, and two and a half hours to return.

We followed the Lahn River through luxuriant mountain country where villages snuggled under cloaks of timber, anchored now and then by high imposing castles. Everything was multicolored. In the green fields grew symmetric patches of poppies and daisies, their sparkling, dazzling reds, yellows and lavenders shimmering in sunlight. Cherry trees loaded with rich, red fruit. grew in gardens or wild along the road. I tried to jump out a couple of times to grasp a handful, much as I love cherries and fresh ones at that, but the boys grabbed my shirttail just in time.

We hit the Rhine at Nieder-Lahnstein and drove downstream to Coblenz. Here we got out and walked across the river on a pontoon bridge. Coblenz itself wasn't in a complete shambles or utter ruin. There were still a few walls left standing. At Remagen we scrambled over the Luddendorf Bridge, the famous one the Krauts had failed to destroy and allowed us to capture intact. That shortened the war by several months. After our troops had gone across and were pushing into Deutschland, it collapsed from strain, killing some workers who were on it at the time. Massive, twisted girders jutted out of the river or lay on the bank as if they'd always rested in those distorted positions. The towers on the bank were strewn inside with abandoned Nazi equipment like clothing, paper and ammunition. A railway tunnel at the end of the bridge was blocked with a screen and a sign warning of mines. Two freshly painted signs had been placed on the abutment in memory of "those who fell when this bridge collapsed." I picked up a misshapen rivet as a meager souvenir and after getting our few pictures out of the way, we continued on.

At the Ittenbach Military Cemetery all was quiet. The Stars and Stripes rose above rows of white crosses. On the front of each cross was a name and a number, and on the back, a dog tag. We walked along the lonely, gravel paths amid the real cost of war.

Remagen Bridge on the Rhine.

Shortly after turning away from the Rhine, and while yet many miles from Cologne, we could see the two majestic towers of the famous cathedral, rising above the hills and fields, hazy in the distance. Driving into the outskirts of the east-river side of the town, we ran alongside miles and miles of wreckage. When we got down to the river and looked across to the heart of the city we saw that the ruin was maximum. I couldn't ask anyone to imagine such total ruin for I didn't believe my own eyes.

The two main bridges from the city across the Rhine had been smashed. We crossed the Gen. Lesley McNair Bridge, an improvised frame structure, and tried three different ways to get to the cathedral, which was standing alone amid vast piles of brick and pulverized rock and cement. So profuse was the debris that we came up against dead ends of streets choked with fallen brick and steel, and soon found that only certain routes were open through the city, and we'd better stay on them if we knew what was good for us. No building was left except the cathedral, no evidence of civilization but for an occasional wall still upright. That was the once-proud city of Cologne.

As we drove up to the cathedral, the height of it stunned us. So did the intricate and complex design with which it had been constructed. Hundreds upon hundreds of little sculptured frescoes adorned all parts of the outside to dizzying heights—each depicting some phase of Biblical history. It gave the impression that construction so many years back in history was impossible, yet a tribute to early architects and builders. Birds flew among the many towers, and their songs resounded with a hollow echo down through the portals. Outside on the steps, civilians sold souvenirs for cigarettes.

Cathedral of Cologne.

Allied precision bombing was not so good that it didn't hit the cathedral, for several bombs had landed on the roof. Bricks, glass and plaster had come cascading down several hundred feet to lie on the ornate floor about the center of the church. There were sandbags placed in many spots, even up high on the pillars, at almost inaccessible places. In other corners were huge smashed and broken wooden girders and more debris. Dirt and dust covered everything. In one corner we found a glass cupboard containing dust-covered shelves of little cloth boxes. One lay on a pile of wood, broken open to reveal human bones within.

Such was Cologne and its cathedral, awe-inspiring, at once indicative of human capacity to build and destroy.

> *COLOGNE, the city of two thousand years, on the Rhine! What a wealth of history lies in the sound of your name. The camp of the Romans, through which the step of marching legions echoes. The city of the Franks, who founded your fame anew. The proud city of the German emperors, for whom you had both shelter and defiance. The home of enterprising citizens, whose brain and industry spanned the markets of the world. Does not the magnificence of the lofty Cathedral towers and the massive structure of your gates give ample proof of your great past? Is it not their spirit that is still to be felt in the venerable Hansa Room in the City Hall, in the halls*

> *of the Gurzenich Building, and in the pictures of the Cologne School in your museums? Moreover, you are the City of Gaiety; and that side of you, above all, is what your visitors keep in memory. City of the present day! You have not surrendered the heritage of many centuries. The forces that you derive from it, you increase a thousandfold. You make use of them for this day and for the future.* Qui non vidit Coloniam, non vidit Germaniam.

If we'd had time, we would have gone over to Aachen, but after visiting the cathedral and a couple of statues which were still partly in place, we returned to Dietz by way of the Autobahn, a super highway that made for speed and smooth riding.

At one place, a massive brick and concrete bridge had been destroyed and had fallen into a canyon, damming the river and creating an artificial lake almost a mile long, submerging several farmhouses and a railroad. But then, the railroad was probably not running for other reasons farther on down the line. The only one we had seen in operation was back at Assmannshausen, across the river, where several freight trains were seen and once a passenger train that had people hanging all over the sides of it. Otherwise, the squealing train whistle was a rare sound in defeated Deutschland.

That, in effect, ended our experiences on the Rhine. We wanted more, but it would not be. Our sightseeing, and indeed our whole period of occupation, was brought to an abrupt and screeching halt that left us dizzy. For it came with such a sudden, deafening blow that coherent thoughts were left out of the picture. But then, they say, things happen fast in war and you can never tell what to expect. How right that was, we certainly found out. Like nobody's business.

THIRTY-EIGHT
LOOK HOMEWARD

The Eighth Infantry Division was re-activated on the first of July, 1940, at Fort Jackson, South Carolina. Early in 1942, just after Pearl Harbor, the Division was assigned the task of patrolling the Atlantic coast, which it did from South Carolina to the Florida Keys. In the middle of 1943 the 8th went to Camp Laguna, Arizona, for desert training. It spent time in Camp Forrest, Tennessee, and in Fort Leonard Wood, Missouri, until the end of 1943, when the Division sailed from New York to Ireland.

On July 1, 1944, a convoy of four troop ships and 12 motor transports steamed out of Belfast Harbor, carrying the 8th Division to the Continent. It assembled at Cherbourg and launched into battle early in the morning of July 8th. Came the Ay River, Brest and the Crozon Peninsula. At one point a lieutenant, leading his platoon through knolls and emplacements, came upon some Germans waving white flags. A German medical officer announced in perfect English that General Herman Ramcke was in a dugout below and would like to talk terms with the American commanding officer. Ramcke, Commander of the Port of Brest until its surrender a few days previous, was rumored to have fled to the Crozon Peninsula. Brig. Gen. Canham, Assistant Division Commander, and his assistants, arrived at the dugout. They were escorted down a concrete stairway about 75 feet underground, where General Ramcke was waiting.

The Nazi commander addressed General Canham through his interpreter: "I am to surrender to you. Let me see your credentials."

Canham pointed to the doughboys outside and said: "These are my credentials."

The 8th broke the Siegfried Line in Luxembourg. Then came heavy winter in the Hurtgen Forest, pain from the Ardennes breakthrough, then the drive on Obermaubach. In storming from the Roer to the Rhine, the Division crashed into Duren, then slammed to Cologne. The 8th was active in the Ruhr Pocket, then lashed across the Elbe River and ended up near Schwerin when final victory was achieved. Like many other divisions it took up occupation duties and in the first week of June some of the Division was near Kassel.

I bring this up about the 8th Division for a good reason. I was soon in it.

The day of June 12, 1945, dawned just as peacefully as almost any other day during the year and there was no premonition of anything unusual about to occur. Nothing to hope for. Nothing to get distressed about. Nothing to worry over. We waked, went down the mess hall and shoveled down breakfast, then came back to police up the room, tidy up a bit as we always did. It was the normal routine. I don't recall what we did during the morning, probably because it was inconsequential, maybe a class in electricity, maybe a swim...

Dinner came and went in the usual way. After noon, I decided I wouldn't lie down and go to sleep for a change—I'd write a letter. So I sat down and began writing to Aunt Florence... dreaming of her pies... her smiles... the soft voice saying "There is nothing we can do to repay you boys for what you have done for us."

It was then that the whole thing happened.

Someone poked his head in the room and yelled: "Sutton! Get packed. You're going home." Then left.

That is the kind of stuff that results in disbelief, fainting, and heart attacks. And it was this, I think, that planted the seed for Enduring Lesson Number Ten: *When you get completely bowled over by something in life, get up.*

ENDURING LESSON #10

"When you get completely bowled over by something in life, get up."

I rushed out into the hallway and chased down the facts. Twenty of the young squirts in the company—those whose point count didn't individually go up much above 30—had been selected for transfer to the 8th Signal Company, 8th Infantry Division. This Division, we were told, was just about ready to depart for the good old USA.

Wham! The bottom dropped out of the 570th and we began to dance with one another in glee, while a lot of glum faces looked on in envy, faces of lost youth, of lost hopes, of gladness for our happiness. There was an outpouring of all the mixed reactions that occur when someone else gets to do the one thing you've had your heart and mind set on doing yourself for so many months or years.

When we finished dancing, we packed our duffel bags and bedrolls hurriedly and began making farewells. Soldiers are nothing if not able to move fast. But these farewells were the most unpleasant duty that goes along with making a big change. We shook hands and passed around pictures for keepsakes.

Fred Larsen had forever called me "Que Alto!" since the episode when Madame Joan, near Marseilles, opened the door, looked at my belt buckle and lifted her gaze, saying in Spanish "How tall!" When the news came, he said, forlornly, "Who'll I speak Spanish with now?"

Dry tears of sadness flowed like wine, for we were leaving the boys with whom we'd won the war—the 570th signaleers, our bosom buddies and close pals whom we might never see again. After all, when you've shared a pup tent and shivered in icy winter on the Riviera with a guy named Dody Ratigan, you don't soon forget such a friend.

I cried on each shoulder in turn and said: "Aw, fellas, this is too sad. I don't want to go home. I want to stay here." They kicked me out.

And that was it. Here today, gone tomorrow. Just like that. We went up to the 70th Division CP in the Castle of Orange and waited for the 8th Division convoy to arrive. Impatiently, we sat and talked, alternately pinching each other just in case it was a cruel dream, still not actually believing this. It had come as such a shock that we had to draw all our impressions and dreams of going home together again and give rebirth to new hopes and ideas.

Here it was at last. We'd only earned one overseas stripe but we felt the stinging truth all too well. We might soon be earning another across the wide Pacific. For there was still a war on in the world, and it hadn't been won yet.

The convoy arrived in late evening and we loaded our equipment and ourselves aboard. After a brief wait, we rode out of Oranienstein and left southern Germany behind. Not much can be said of the ride that night. We were crowded like sardines with all our gear—including my extra pair of combat boots. Someone had spilled gasoline in the back of the truck and this stench became overwhelming. Eventually, though, it evaporated and we went to sleep.

We arrived next morning during the dawn hours at a place somewhere near Kassel. We were in a farmyard. We climbed out and went into a low-ceilinged barn, where we parked ourselves and our gear and sat down groggy and sleepy-eyed to wait further action. There were K-rations for breakfast, then sleep most of the morning. For some reason, winter returned briefly. It was cold, bitterly so, with an overcast outside and a frigid wind that crept through cracks in broken windows. But we slept.

At noon, more rations for dinner, and shortly afterwards a brief ride to a small village where the 8th Signal Company was quartered. We unloaded, found our room and spread out our sleeping bags, then spent the evening enjoying good chow and getting acquainted with new friends in a new outfit.

We didn't stay long at this village "somewhere near Kassel," for after two days of increasingly cloudy and drizzling weather, we loaded up in trucks and began the trip to France.

It took three days to make the trip. The first one was spent in preparations, in formation receiving instructions, and then through Cologne and Aachen and down near a place called Bonn. All day long we saw nothing but devastation, ruin, shambles—a shattered country in a shattered world. We camped on a deserted airstrip near a P-38 aircraft graveyard, making 4-man tents to hurry the bivouac process in the rain that had fallen intermittently all day and still came down in curtains.

The excitement and expectations of the trip were beginning to tone down now, and we tired of riding all day. That night we slept like logs.

The second day's ride led us out of Germany, across Luxembourg, past the border and back into France. A few of us made the mistake of standing up much of the way, faces to the west, receiving the full blast of wind and dust for miles and miles. For that, we received beet-colored, wind-burned faces, serving us right not to know better.

We passed more ruined towns and villages, enough to make us wonder how Europe could ever get on its feet again. One burg not too badly scarred was Luxembourg City. This place stands out in memory because of a French WAC we saw as we rode slowly up an inclined street. She walked along the sidewalk with a lady friend, and seemed at the time to have been the most beautiful face we had ever laid eyes on. There was something about the face and the form trimly attired in uniform that left with most of us an indelible impression, a bright and cheerful spot of humanity, you might say, in a world so thoroughly ruined.

Back in France again, and into legal fraternization, we began throwing things to the people who stood and watched us go by. We threw them gum and cigarettes and candy, any little thing we could find. They threw us kisses in return. In some places, I wanted to climb out so that the lovely young mademoiselles would not have to throw their kisses so far.

The second night we spent in professional accommodation at US Army Bivouac No. 1 near Montmedy. Here there was a permanent mess hall, a field wash house and even a canteen that sold beer and Cokes and a few other knick-knacks. The sun hadn't yet set when we arrived, so a couple of us took a walk up the hill across a grassy pasture that was marked *Privat*. We talked for a bit with a Frenchman who didn't necessarily see eye to eye with us on political matters.

On the third day's ride, we wore eye shields that protected our faces from the erosive action of wind and dust. The shields did not, however, cover the nose, so by the end of the ride, I was looking like an elongated, overstuffed version of W.C. Fields after a hefty bender. The nose looked so bad, sitting there between my eyes, that I felt it could peel off at any time.

We went through Rheims and got a fine view of the famous cathedral. After Soissons and Beauvais we arrived at the end—the temporary end—of our journey, the quaint French village of Yerville.

Here we found a cow pasture turned Army camp and bumped around from tent to tent before we found quarters and eventually got lined up with the radio section. It was Sunday the 17th of June. Some boys, with time on their hands and love in view, went out, as they called it, "cat hunting." (With remarkable success, we learned later— providing you got in line.)

In the days that followed, we had nothing to do but wait and speculate and get everything in readiness for the trip. We weren't far from Le Havre now, and knew that was to be our port of embarkation. We turned in much of our excess equipment and I, hallelujah, got rid of that extra pair of combat boots that—and you may find this hard to believe—I never got the chance to wear. If I'd known then what I knew here, I might have dumped those boots in the Atlantic. We got our ETO jackets in shape and packed our duffel bags scientifically, stowing away all the souvenirs, and sat around to wait.

Spring returned, and we spent time basking in the sun and reading. At a department store in one corner of the field some of the troops got rid of their "liberated" binoculars and other equipment that was taboo for taking home. The French had things to offer in return, so quite a lucrative business went on. Now and then, a glance would be made toward the surrounding area to see that no police approached to break up this little black market bazaar.

We enjoyed good food and even had movies for entertainment. We roamed around the village a little, and I brushed up on my French. Some day, I vowed, I'd take a course in French and come back. Several of us visited a house where we attempted to learn an English version of "Alouette" [a round of folk singing], but someone should have told us how impossible that was.

The first word we had of an embarkation date was on Sunday the 24th of June, when word came that we could expect to leave on the 28th or 29th. The latter proved to be correct, for on that evening we loaded with all our equipment aboard a convoy of cattle cars and drove to Le Havre. The evening was overcast and made all the buildings and ruins of the city appear drab and gloomy. But this atmosphere applied only to the port city, for within our souls lay anything but gloom. We were on the way home. We went aboard ship at 0100.

Good fortune did not smile upon us quite as much as on the voyage to Marseilles, for now we did not gain the comforts of a converted luxury liner, but the questionable luxury of a liberty ship, a troop transport numbered P-130 for security reasons and named the General Squier.

At 0600 the ship got under way and steamed out of the harbor at Le Havre, a harbor that bore considerable resemblance to that of Marseilles. Here, too, there were sunken ships and devastation on the docks. A grey dawn came to the coast as it faded into the distance. The light was still gray when the last vestige of land disappeared. With it disappeared the continent that had been our home for the last six months, but the memories would not fade as easily as the shoreline. We would always carry Marseilles, St. Jean-Rohrbach, Merlenbach, Kreuznach, Frankfurt and Dietz with us. I would retain the fond memories of the people, like Madame Joan, the Crochovalskys, and the lovely Taya. There would ever be the vivid pictorial recollection of Provence in winter, Northern France in spring, and the Rhineland; of villages and their streets, of houses, wire, radio and combat. It had been a pleasant enough trip overall, a tour that would have required a small fortune at other times and other conditions. We had done the Continent at an early age.

All day long on the 30th we watched seacoasts to the north of us, far in the distance. These we assumed to be the White Cliffs, at least the coast of Britain. That afternoon, more shoreline came into view, and this we thought might probably be the Irish coast. After that, nothing but sea, ship, and sky.

The trip to the USA went smoothly. The ship was packed to the gills with returning veterans, some en route to a discharge, others like us preparing to enjoy a furlough and then on to the Pacific. There were troops everywhere. You couldn't walk out on deck without stepping on someone or tripping and sprawling over several more. We didn't do much sleeping down in the hold, for July was too hot, and it was nice to curl up in the sleeping bag out on deck, listening to the roar of the waves and the wind in the mast. Nice, that is, provided it didn't rain.

Chow was served in the usual twice-a-day custom and somehow, not so many men got seasick this time. Perhaps that was because the first few days out were calm. As a matter of fact, the whole trip was without bad weather. The ship's PX opened once a day so that we might order ice cream, candy and other items we had so long done without. But we had to earn what you got, for the PX line extended all the way around the ship twice, so that we'd frequently meet friends coming and going. Standing in line like that gave us something to do, however, and that was at least a small advantage. The rest of the time we devoted to playing poker, bridge and pinochle, and to standing at the rail, gazing into the sea, shooting the bull, or napping.

There was considerable sea life, and the first whale brought everyone topside. It followed us along for some time, coming out of the water like a surfacing submarine and then slipping slowly back into the depths, spouting a geyser as it did so. Its black tail swept out of the water, made a graceful arc, and slapped back into the water with a smash and a spray of water. There were porpoises, too, leaping in the sun and oftentimes following closely alongside the ship. Occasionally they were attracted by garbage dumped from the ship's mess compartment. We couldn't see any sharks, but thought it natural that some ought to be around, with that much surplus food going overboard. We studied the emerald water close to the ship and many times saw smaller fish swimming about, fish approaching two or three feet in length and about which we knew nothing.

On the 4th of July the ship's crew shot varicolored flares into the air and fired salutes in celebration. A couple days later someone yelled **"MAN OVERBOARD!!"** Cards, blankets and dollar bills flew as we stormed up to the rail for a view of operations. A voice came sharply over the loudspeaker system for everyone to stand back from the rail so that rescue boats could be lowered without congestion. But no one moved.

They got the boats down, however, almost. The one nearest which we stood was manned immediately and never did get over the side—some technical difficulty. We kept looking astern to pick up the form of the man who had fallen in, hoping he wouldn't be pulled under by the suction of the propellers.

Someone yelled: "I see 'im!" We looked long and hard, but the situation then became ludicrous when the voice over the loudspeaker announced that the drill was complete and for all ship's personnel to return to their stations and continue about their regular duties. Considering their efficiency of action, I decided it would not be wise to jump overboard and expect to be saved.

On the 7th of July we began to perceive debris floating on the surface—boxes and trash typical of mainland refuse. Then came gulls zooming around the ship. We strained our eyes long toward the western horizon but couldn't make out the faintest traces of shore. We saw a lot of imaginary coastline, just because we wanted so badly for it to be there, but finally gave it up and went back to playing cards.

Later came the cry **"LAND, HO!"** Cards and blankets flew again as everyone crowded the rail. Sure enough: once more we were feasting our eyes on the Promised Land.

It was an emotional experience, to say the least. Here was the coast of Virginia. We saw the beach, then the trees and a few buildings and lighthouses, and finally, as we drew closer, the city of Norfolk. We scarcely blinked our eyes during the arrival, for fear we'd miss some of it. We saw a sight to which we were entirely unaccustomed. There were large buildings and smoke-stacks, all intact! The ships in the harbor were afloat, not sunk, and the docks bustled with activity. Everything was green and lush; we saw parking areas and thousands of cars gleaming in the sun. There were shore driveways where automobiles came and went.

As the rays of the late afternoon sun shone down on the harbor, we came to a stop. Shortly afterward we began to move again, the theory apparently being to back up and turn around so as to anchor properly. Then the ship ran aground.

By this time, we had almost resolved to swim ashore if that would hasten our debarkation. The USA was here and we weren't about to let it go, not this close. The ship got loose at nightfall, but remained out in the harbor during the night.

We didn't sleep much that night. A few of us placed ourselves in a gun pit and snatched a few winks, hoping not to miss the docking. Everyone was up and ready at 0100 but still nothing happened. As dawn began, we read the signal communications from ship to shore. We could only see the faint flash of our ship's lamp on the buildings on shore. After sending a couple signals and getting no answer the ship turned on what must have been an arc light, which was enough to awaken anyone on shore. It did. Came the answer: "Proceed to anchor. Troops must be fed by 0530."

It was too early in the morning for any of the populace of surrounding towns to be down at the wharf to give the boys a rousing welcome, or perhaps the arrival was a secret. At least, there was a military band hitting up the music for our arrival, and a few WACs circulating around for atmosphere. It took most of the morning to unload the ship and all its eager personnel. I'll not forget prancing down that gangplank and touching American soil (or dock boards, it's all the same) and feeling how wonderful it was to be back again in the land of plenty.

We waited for a while in the pier house and eventually boarded a train for Camp Patrick Henry, just inland from Norfolk. Here we enjoyed the luxury of real, honest-to-blazes military barracks. Yet even if we enjoyed this, it was nothing to prepare us for the meal they served us, the first real one once again on home soil. It was big and had everything from chicken to ice cream. We ate and ate and I thought a number of the explosions I heard were bursting tummies. Mine I knew would go off at any minute, but what the hell. I couldn't stop if I'd wanted to, and I sure didn't want to.

Eventually, though, I forced myself to quit, and staggered outside and back to the barracks, to spend the afternoon recuperating. What a heavenly meal! Few others have been so enjoyable, but then, no others had ever been eaten under such circumstances. Even a short period of six months abroad can do that to you. We had tremendous sympathy for the guys who were in Europe for years...

At Camp Patrick Henry, the heat and humidity made us feel like hot wet towels that hadn't been wrung out. It was detestable, but it didn't last long, for the boys were soon dispersed to their various reception centers. I went through Virginia, West Virginia, Kentucky, Indiana, Ohio, Missouri, down through Arkansas and into Texas, from chair car to troop sleeper. The train was well marked with expressions from the Continent and souvenir Nazi flags draped on the sides of the cars to let people know beyond question just what kind of a troop train it was.

Clear across Texas and into the Logan Heights Separation Center at Ft. Bliss, just north of El Paso. Here we received new clothes, fresh clean khakis that made us feel that once again we were leading a civilized life. A short stay here to clear up details and arrange papers, and then, at last, to that temporary destination:

The Empress Mariposa slid along the bay
Slow but getting faster, taking them away;
Boston lights behind—so carefree, so gay,
Came down by the water's edge to play.

They went up on the deck to think—
From out the darkness came a rapid blink:
"What ship?" "Troop Transport!" Do not sink!
Then went below to catch a wink.

They waked up then in half a year,
And saw the waters not so clear:
Behold! They saw a thousand cheer,
Off port the Dixie coast was here.

What is this thing called overseas?
What does it mean to stand and freeze?
Means nothing! For what he sees
Is something standing 'neath the trees.

His glorious deeds contained in many a tome;
He didn't care so much the world to roam,
Wanted merely just to cross that foam
Back to something he called... home.

EPILOGUE

Shorty went on for another thirty pages, but a summary here will do. While he was home in Arizona, a big flash of light lit up the sky in New Mexico, and the next morning, newspapers were full of speculation that a meteor had crashed into the White Sands near Alamogordo. Later a blast of light and energy lit up the sky above Hiroshima, Japan. Then Nagasaki. And the War in the Pacific ended. How many lives those blasts destroyed—and how many they saved.

By sheer coincidence, he found himself back in Fort Leonard Wood, Missouri, to occupy the same barracks with the 8th Division Signal Company. It was a time of gross uncertainty, counting the points that made you eligible or ineligible for discharge. Much of it was an easy life, but Shorty, impatient, wanted to get back to civilian life and start college.

No way. A short while later he found himself with the Second Division Signal Company at Camp Swift, in the scrub-oak center of Texas. There he ran the company radio school for a while. He also added another to his long list of youthful experiences. He went into surgery at the camp hospital. A sebaceous cyst had formed a lump on the back of his neck, and so they strapped him on a surgical table, put his arm out on an extension, and tied gauze around it.

"That's not going to hold me down," he scoffed.

Two massive arms of an obviously champion weight-lifting orderly clamped down on his wrist. "*This* will hold you down," the man said. When the nurse said: "Count to twenty," and jammed in the needle with sodium pentothal, he only got to twelve.

He spent weeks in the hospital, and they never did get all the cyst out, but it never bothered him again.

They transferred him to Fort Lewis, Washington, and en route he joined masses of troops for a parade down Market Street in San Francisco. The last of his Army days in 1946 were spent among the northwest's hemlock forests beneath Mount Rainier. He was finally discharged on May 6 in Jefferson Barracks, Missouri.

However, he made one mistake. He joined the reserves.

Nothing happened for four years. He went through college with flying colors and got a degree in biological science. Then Korea happened, and for a year he spent time at the Signal Corps Center in Fort Monmouth, New Jersey, in a top-secret unit developing ways to deceive the enemy on a battlefield using electronic means.

He didn't intend to get back in. He had entitled his World War II book *Once is Enough.* But flexible as ever, able to stave off the jabs life threw at him, guided by his Enduring Lessons, he became philosophical. The last phrase of his military writings came very easy to him, and he meant it from the bottom of his heart.

"Come what may, there is one unquestionable fact. The US is worth fighting for."

❦ ❦ ❦ ❦ ❦

ABOUT THE AUTHOR

Myron Sutton was raised in Arizona and graduated from Northern Arizona University in the field of botany. He and his wife, Ann, a graduate geologist, met when they were ranger naturalists at Grand Canyon National Park. He served with the National Park Service, of the U.S. Department of the Interior, for 30 years, the last ten years as a consultant to 80 foreign governments.

Together, Dr. and Mrs. Sutton have written and published 23 nonfiction books on nature and the natural sciences, including *The Appalachian Trail, Yellowstone: A Century of the Wilderness Idea, and Eastern Forests.* Recently they embarked on a series of mystery novels set in world heritage sites. Published under Ann's maiden name, Ann Livesay, the books are geographic and ecologic suspense stories in wild and dangerous places.

As professional photographers, their world images are marketed internationally by agents in New York and abroad. They live in Medford, Oregon.

A BARRY ROSS International Mystery

Take a walk on the wild side with these exciting excerpts from the first five Barry Ross International Mysteries. To order, send check or money order for $US12.95 per copy, postpaid, or send Visa or Mastercard number, expiration date, name and address to Silver River, Inc., 1619 Meadowview Drive, Medford, OR 97504. Or visit our web site at www.silverriver.com.

THE ISIS COMMAND

(Barry goes to Egypt to tackle a syndicate selling artifacts robbed from ancient tombs.)

My mind whirled with Apis's revelations. I was hopelessly trapped. There was no use in putting out an all points call to board every felucca in Cairo and find the one laden with explosives. The martyrs would blow up the Cairo waterfront if nothing else. I had to find out which felucca had been earmarked for a death cruise and let it get out into the Nile before we slammed a missile into it.

We? I needed help. The unguarded tombs burned a hole in my mind. The next job was to find out where to get help. Maybe from the Navy. It was a long shot. But I was used to banking on long shots.

Then, surprise! At the hotel, I ran into the Swains.

"Barry!" Veronica gushed. "Praise Allah! We've found you."

"What's up?" I asked.

They took me by the arm and led me into the restaurant. Over a cup of tea they made their momentous announcement.

"We had a meeting," Tab said. "All of Maybelle's mob. They decided they were going to help. It was unfair for you to shoulder this burden of Turki's murder alone. It's not just your job. It's our job. All of us. All of Egypt. We've got to find out why he was killed and who did it, and stop this kind of stuff."

"We know why he was murdered," I said.

That visibly shook them. They looked at each other with wide eyes. "Why?"

"Tell the archeologists that I deeply appreciate and admire what they have done. Tell them also that they are not trained investigators. This is a very dangerous case, and I do not want to endanger them. Time is very short. I have six problems in my lap and they could blow up at any moment. Literally."

Veronica held up her hand. "I knew it. That's why we want in on this. Which problems shall we take?"

Tab added: "We are not going to accept any nonsense, Barry. You have been overruled on this. Surely there are some things we can take care of and leave all the bang-bang-shoot-em-up to you."

"I won't have it." As soon as I said that, I suddenly realized that maybe he had a valid point. I didn't want them underfoot. If I could sick them off on some wild tangent, they wouldn't interfere with what I had to do.

"You can't stop it," Tab said. "You might as well be reasonable and bring us into this."

"I won't."

"If you don't, we'll start boring into this thing ourselves. That'll screw you up more than anything else."

"You don't know where to begin."

"That won't stop us. We'll find out. The mob is absolutely determined about this. You saw them. You met them. They're a bunch of stubborn characters and you know it. You ain't gonna stop them."

"Who's behind this?"

"Sandy, Hami, Ali, Molly, Mei Hua, and us. That ought to scare you!"

"It sure as hell does. What about Maybelle?"

"She doesn't know anything about this. She'd only disapprove it."

"Well, there's one sensible head. I told you this was dangerous, and now I know that it is more so than I thought. Deadly, in fact. And I'm not surprised."

"How? What do you mean?"

"Never mind. The whole idea is absurd."

"Where's your friend Samaan."

"In the south."

Veronica squealed. "You see? You can't go on with this alone."

I stopped and thought it over again. This wasn't going to quit or go away. I could see that. The last thing I wanted was a bunch of high-grade, hot-headed scientists diving off into deep water. They could botch up everything I'd been trying to set up.

They didn't know about the tombs Turki had discovered.

They didn't know about Zebec and the syndicate. At least, I didn't think they knew. They didn't know who Isis was, the master mind. Or maybe they did. The situation was confusing enough without all this complication.

On the other hand, they were all pretty intelligent people. But if I sent them off in different directions, they could flounder all over the place, stumble into danger, and interfere with what I was trying to do.

When I finished arguing with myself on this, I decided that it would probably be better to put them to work. My only hope was to get them out of the way long enough to choose my own path.

"All right," I said. "I give up. You win. But I want it clearly understood that I am in charge of this investigation, and if I give somebody something to do, I want it done according to specific instructions. Is that clear?"

Veronica looked at Tab, stunned by the quick turn of events. Then back at me: "Why, of course it is. Nobody doubts you're in charge. They want to help. They just want instructions."

"All right, then, I'll give them to you."

I brought them up to date about Turki's discovery, and I thought for a moment that they were going to fall off their chairs. More astonishing, they were speechless for all of twenty seconds.

Tab said: "How in the hell did you find all that out?"

"Because he's smart and you're not, you dumb ass," Veronica said.

"Never mind," I responded. "The first priority is to get that tomb sealed and guarded. Omar's gone, and the head of antiquities is in Greece. I don't want to trust this to anyone else. I'm going to see if I can get some American military help."

"American?" Veronica exclaimed. "Bully for you, Barry!"

"I do not, repeat, do not for a minute expect success."

Veronica's face fell. "Stop being so negative!"

I told them to get out a notebook. An archeologist is never without a notebook. They did. "Now write this down carefully. I'll be depending on you to transmit these instructions without error. Okay?"

"Okay," Tab answered, getting out an old, battered, aluminum-bound field notebook.

"Tell Hami to look up anything he can find about syndicates in Cairo. The names I've run into are Zebec and Apis, but those have to be false names. Have him do what he can. Tell Ali Fawzi to plead with someone in Turki's office for help in guarding the newly discovered tomb.

"I'll try to get some temporary U.S. military help. It's going to be tough, because things are destabilizing here and there, like the militant bombings in the south. But tell him to keep at it. Ask Molly and Mei Hua to trace down an underworld character named Isis. This person, according to Apis, is the head of the syndicate, and may be responsible for other tomb robberies. I want this whole thing stopped, and it can't be done without snagging the top brass. Clear?"

"Clear."

"Try to get Sandy to take all his equipment to the new tombs and make a complete record, trying not to touch anything. This will be tough. You'll have to get a helicopter pilot named Amir to take him out there. I'll pay all costs if the Egyptian government can't or won't. I want full visual and digital records. Then he can go on back to Thebes and finish his work."

"What about us?"

This was my toughest part. If I didn't give them the most important task, they would be miffed. "All right," I said, "you get the most dangerous part."

Veronica clapped her hands. "Oh, goody!"

I told them about Apis's revelations, that somewhere on the waterfront was a felucca loaded with high explosives. I wanted to know which felucca it was, what kind of explosives it had on board, when it was scheduled to sail, and where it was going.

"That's impossible," Tab said, in a wilted voice. "No one talks in Egypt. Not about things like ***that***. Jesus, man!"

"Apis did. But I don't want you to go around asking a lot of questions. Just observe. Do you get me?"

"Yes. But we can't answer many of those questions by just observing."

"Do the best you can. And make this quick. See if you notice any felucca with a big hidden, tied-up cargo, and make a note on any distinguishing characteristics. When it takes off, we have to follow it down the Nile..."

"How?"

"I don't know. Get out the maps. Have you been out on the delta?"

"Sure."

"Check out the roads closest to the Nile. My plan is to keep track of that felucca, follow its progress. You got your Land Rover?"

"Yes."

"I'm assuming they will sail down the Nile and out into the Mediterranean. In this country there isn't much other choice. My job is to get in touch with the Navy."

"How will you do that?"

"Never mind. If you have any messages, leave them at my hotel. You have a cell phone?"

"Yes."

"Okay, we'll try to synchronize on that."

"Anything else?"

"Yes, my two assistants are here on a mission to buy the artifacts from the tomb and help trap the syndicate gang."

"Names?"

"Julie and Joe Muck. I'm going to call them now. But no one is to mention them. They are supposed to be innocent buyers with a lot of cash. I mention this to you because they may be involved very deeply in what happens. See you later."

They rose and rushed in great excitement out of the room.

I pulled out my phone and tried to call Julie and Joe. No answer. I tried the Club Med desk. *Les Mucks ne sont pas ici.* Lapsing into French, I asked where they'd gone. The response: some men came to see them, and they all went away together. What kind of men? I asked. One of them had long hair tied in a knot behind the back.

How long ago? An hour.

Dropping everything, I got out to Club Med as fast as I could. No Julie or Joe. Baggage still in their room. No notes. No phone messages. They had not had time to tell me—or anyone—where they were going.

I called Omar in southern Egypt. Nothing in the police log. Besides, Omar was so far from Cairo . . .

My worst fears had been realized. At first, I tried to tell myself that Julie and Joe had just gone to the little shack to make a down payment on the artifacts. Maybe it wasn't as ominous as I thought.

I was only kidding myself.

Zebec had Julie and Joe.

(Available now)

DEATH IN THE AMAZON

(Barry leads a dugout canoe trip down the Orteguaza River in the Colombian Amazon basin.)

Boris was in good form. "One day this will all be opened up. All of it. People won't starve any more. You're out of date, sweetie. The forest was made to be *used*. Anybody knows that. Why do you think it was put here? We can ship out the logs, pump the oil, mine the gold...."

At that point, Joe Muck, sitting beside me, had taken all he could. "Beautiful! Just beautiful! The rivers are already running with mud and mercury. Who the hell do you think would *want* to live here? If they did, they'd choke. They'd starve. What would you do with the tapirs and caimans?"

Boris answered in a flash. "Sell 'em! Put 'em to use! Turn 'em into handbags. Make belts out of 'em. Haven't you seen crocodile boots in the stores? Get with the fashion, boy!"

That was all Mo Briggle could suffer. He turned. "What kind of perverted viewpoint is that?"

The Arden-Joneses turned. Chris scowled. "Is that chap real?"

Spider outshouted them all. "Where are the piranhas when we need them? You idiot! You're blind as well as dumb. Every law in every country is pointed in the *other* direction. *You're* the one who's obsolete."

A babble of angry voices erupted—only to be cut off by a flash of lightning and deafening clap of thunder. Dana Briggle shrieked and buried her head again.

Joe Muck shouted: "There's your answer, Boris! Listen to that! The gods are after you!"

"They're on my side!" Kolinski shouted. "They're on the peoples' side!"

"Bullshit!" Joe shouted above the tumult. "They've come to take you out. I just hope they don't take us all out at the same time!"

From the black sky came another bolt of lightning and a deafening thunderclap.

The storm had arrived.

Nothing could have so effectively drowned out the growing animosity between Boris and his fellow passengers. The "rain" in rain forest here is not the gentle drizzle that drifts into a temperate rain forest. Here the water falls by the ton—bashing, thrashing, battering and drowning.

A wall of heavy rain bore down upon us from behind, pushing violent gusts before it. The canvas canopy began to billow and thump, rocking the canoe. We clutched the edges to hold ourselves steady, but it was hopeless. Gusts of roaring wind threw us from side to side. Dana buried her head again.

The Indian pointed to the right. Eduardo turned the vessel and guided it toward shore.

José found an eddy within a sheltered cove and took refuge. Eduardo gripped the limb of a fallen tree, hoping to keep the canoe in place—a pitiful effort at best. The furious wind wrenched the limb from his hand. The canoe skewed sideways, end pointed to shore.

It was a perilous moment, in which the canopied vessel, broadside to the winds, could have been blown over and all of us toppled into the whipping waves.

The Indian scout plunged into the water and swam ashore with a rope, which he lashed to a fallen trunk, securing the front end. Then he swam back and resumed his place on the prow.

Monty leaped into the water, clothes and all, and tied the stern to another tree limb. It was a precarious hold, but he had no choice. He hopped back aboard, just in time.

A powerful wind nearly blew the canoe against the bank. A liquid curtain deluged us like a waterfall. Heavy drops pelted the canvas with a roar. Torrents of rain splashed into the canoe at both ends, drenched the boatman, scout, and Adams, and roiled the river surface. The tropical Orteguasa had become a cauldron of exploding water.

No one could hear anyone else. The Fishers, up front at the edge of the overhanging canvas, recoiled and pulled jackets over their heads to keep out the heavy spray. They tried to shout, but the roar absorbed every decibel.

More lightning flashed, closer than ever. Thunder pounded, echoed, and reverberated, as in a vast celestial amphitheater. Tons of water fell on the sagging canopy and all around us. The roar intensified. The shore, close off the starboard bow, vanished from sight as if we had passed beneath a Niagara-like cascade.

Joe shouted "Whoopee!" but only the din of the rain could be heard.

Water began to flow along the bottom of the canoe beneath our feet. In the back, Adams bailed as fast as he could with a pitifully small can.

The canoe shifted and turned, pulled on the ropes, strained against the limbs to which it had been hastily attached, and tried to break free.

No one could see ahead, to the side, or behind. If a floating tree trunk, driven by wind and wave, came down the river on this side, no one would know until it hit. If the river rose suddenly, pushing a giant wave, we would scarcely have seen it until too late.

Under such a powerful onslaught, we could do nothing but sit, weak and helpless, victims of nature on the rampage. The water in the canoe became deeper, covering our feet. I cursed myself for not having brought a lightweight plastic or canvas bucket to help bail out the water. I should have thought of that.

The pounding pandemonium seemed unstoppable. By a miracle, the canvas held. Under such a downpour, weaker cloth would have ripped to shreds or blown away. If that had happened, the canoe would have rapidly filled and sunk.

The maelstrom moved on downstream. The torrent stopped abruptly, leaving us gasping, our ears ringing. The wall of water receded. The river bank reappeared. Then the river. Then the forest on the other side.

We sat with shoulders hunched, expecting another lightning bolt and tympana of thunder. Only the sound of rain pounding the river came to our ears, and soon this faded.

"Well," Joe said, "*that* was a nice little shower. Are we sinking yet?"

In the back, Monty scraped furiously with the little can.

Mo had covered his wife's head with a jacket. She remained as though buried in a burrow.

The Fishers brought out their cameras to photograph the receding rain, but found the instruments wet and the battery compartments short-circuited. Greg cursed.

"Plastic doesn't do it! Pull over to the nearest supermarket. We need a rubber bag."

Joe said: "Too late, my friend. The stores are closed."

Boris: "When are we going to stop for dinner?

The river and forest absorbed this massive downpour as if nothing had happened. In the aftermath, we expected the river to be in a destructive flood stage. Not a sign of it. A few wisps of mist rose from the surface. Around us, the Orteguasa revealed no more eddies and whirlpools than before.

The cloudburst had been absorbed. Just like that.

The Indian swam ashore, released the rope, and carried it back to the prow. Monty went in again to untether the stern, and climbed back aboard. Eduardo revved up the motor. The canoe moved out again into the open current.

(Now available)

THE MADMAN OF MOUNT EVEREST

(A grisly murder in the peaceful Khumbu Valley beneath Mount Everest sets Barry and Joe off on a seemingly hopeless but hazardous pursuit.)

October 28. Phakding, Nepal. 5:30 p.m.

We spent the night at Phakding, a small settlement on both sides of the Dudh Kosi.

The little inn served what I called "your culinary introduction to life in the Khumbu." A carrot salad, bread chunks, rice, potatoes, brussels sprouts, momos and cheng. Joe particularly fell for the momos, golf-ball size chunks of yak meat boiled in a pasta wrapping. Ari wolfed them down like a starving hiker.

"You two eat any more," I warned, "and we'll have to get yaks to carry you both up to Namche." I thought a moment, then added: "Or back to the hospital in Kathmandu."

When he sampled the cheng, a beer brewed from rice, Joe commented: "They've reinvented the *chicha,*" referring to a native brew of the Amazon Indians. "This will keep your bacteria alive!"

When we had finished, Joe rose: "Now for a nice hot shower."

I responded: "Did you bring along kerosene to heat the water?"

"Huh?"

"No hot showers, Joe. In fact, no showers at all."

Joe sat again, a stupified look on his face. "You're kidding. Yes. You must be."

"Wood is at a premium up here. It's not going to be wasted to heat water for tourists."

Joe frowned. "I'll disintegrate."

"Try the Dudh Kosi. Go for a dip."

"Come on, Barry," he said pitifully. "I'll freeze"

"Okay then," I commiserated. "You'll only get dirty. It won't kill you."

"But..."

"You get dirty on elk hunting trips, don't you?"

"That's different—."

"Cool it, Joe. You're not going to get a hot shower as long as we're up here. So the sooner we capture our prey and return to Kathmandu, the better."

"All right, then, the hell with it. I'll go jump in the river. I'm tough."

"You want to know the temperature of water fresh from glaciers?"

Joe rose, went into the screened-in bedroom at the front of the building, and fell with a "thunk!" on the bed. "Good night, everybody."

The night passed uneasily. We slept in a room for six, on cots with boards instead of mattresses or springs. Joe's lanky frame felt the pain of having only a sleeping bag between him and the boards. He tossed from side to side all night. I know, because I'm a light sleeper, and every time he rolled over, the fragile cot squeaked.

Joe must have spent many a night on pine needles in the Sunlight Basin—without moving a muscle. Not in Phakding. Here, we had a problem.

Two porters occupied beds at the other end of the room. Both, en route to Namche, had carried heavy loads on their shoulders all the way from the lowlands. One bore a sack of salt, the other a sack of rice to sell at the bazaar.

They had smoked heavily in the kitchen all evening, and by bedtime, the rooms reeked with cooking fire smoke and, Joe thought, some of the most acrid, eye-smarting tobacco smoke he had ever experienced. The open, screenless, glassless windows didn't offer much help. The air in this narrow canyon clamped down on us and didn't move.

The porters coughed half the night. Like machine-gun fire, their staccato ***kaboom-karaa-kop-kop-kop!*** erupted in a fit of sneezing, snarling and spitting on the floor.

Kaboom-karaa-kop-kop-kop!

I wondered how any human being could carry heavy sacks for miles and miles all day and cough all night. Eventually, we succumbed, but not until after midnight, after the coughing stopped.

Next morning, the thickening smoke and clamor of kettles in the kitchen waked us. As we dressed, the ***kaboom-karaa-kop-kop-kop!*** began again.

Joe asked softly: "How could you sleep through all that?"

I didn't answer. I glanced at the porters and then addressed Joe discreetly: "The usual custom on returning home, Joe, is to go to your doctor and get a skin test for tuberculosis. You have a doctor in Laramie?"

Joe looked at me sharply. "TB? You're kidding?" He studied my eyes and added: "No. You're not kidding."

Joe shook his head to clear away sleepiness. "Wanna jog before breakfast?"

That always got a positive response from me. Except now.

"Save it, Joe. You're going to need all the energy you can get this afternoon."

Pasang joined us for breakfast. The yak was already loaded and waiting outside.

We received a mixture of scrambled eggs and rice, with nak milk.

"Nak?" Joe asked.

"Female yak," I replied.

Pasang said: "You like, huh?"

Joe's lips curled. "Isn't there any cow's milk?"

"Where," I asked, "do you think they would get cows?"

"Same place they got yaks."

"You're stalling, Joe. Drink it. You'll need it. We have to get going."

Joe cringed, closed his eyes, and drank. Then he gave up his own version of ***Kaboom-karaa-kop-kop-kop!***

After a few minutes, Joe pointed outside and said: "Look at those porters. They're *smoking!* Did you see that? They climb the highest mountains on earth. Oxygen nearly ceases. They carry hundred-pound loads. And now they're smoking. Am I missing something?"

I smiled. "Keep an open mind, Joe."

"And what about the yeti? The abominable snowman. Truth or fiction?"

I smiled again. "Keep another open mind, Joe. Ask Ari. He knows everything."

"He's still asleep. Let's leave it at that!"

With the help of a high protein energy bar and a last sip of sweet Tibetan tea, we roused Ari, loaded our packs, and headed up the trail...

(Now available)

THE CHALA PROJECT:
Murder in the Grand Canyon

(Barry and a squad of park rangers have been assigned to accompany Hounto Chala, a high exec in the United Nations Environment Programme, on a raft trip down the Colorado River through the Grand Canyon. Trouble is, Chala is marked for assassination.)

"Oh, my God!"

Cynthia Kasbolt turned and ran toward Joe and me.

"What's the matter?" I asked.

She pointed toward the river. "Now he's swimming!"

Sure enough, Chala had peeled to his swimsuit and begun to step in the swift-flowing water.

Ted Quail saw him at the same time. "GET HIM OUT OF THERE! THAT'S NOT SWIMMIN' WATER! THERE'S RAPIDS BELOW!"

Joe yelled: "EMINENCE! NON!" and took off.

As he leaped over rabbitbrush he flung off his hat and shirt.

Chala had not heard. Now up to his hips, he began to stagger under the hidden and unexpected force of the current.

"Eminence! Eminence!" Joe shouted.

The African lost his footing and sank to his neck, the torrent hurling him against sharp-edged sandstone layers underwater.

Joe, in one flying leap, pancaked on the surface and churned the waters like a motorboat until he could grasp Chala's arm. The icy water sent sharp pangs through his body.

Chala went under. Joe pulled him up, turned him toward shore. Within seconds, Buck and I hit the water, splashed up and grasped the other arm. By now we flowed along at a brisk pace, nearing the lip of the rapids. The powerful torrent made landing hazardous.

Joe grabbed a rock, spun around, lost hold of Chala. A current surge hurled Buck back. That left me. I lunged with superhuman energy, slammed my foot against a rock and thrust Chala out of the water. A Herculean task—using the leverage of a rock to move a big man against a powerful current. The force of the water threatened to upset a precarious balance and rip us both away to the churning maelstrom just below.

I pushed and thrust with every ounce I had. Within moments we rolled into shallow water, crashing against boulders and slabs of submerged sandstone.

Cynthia Kasbolt and Julie Baxter raced up and helped pull Chala and me from the water. Evie Lund arrived with a towel and led us to the first aid kit. As soon as we had dried off the water and some of the blood, she applied antiseptic salves and bandages to the assorted cuts and scratches.

Julie stormed over to Dietrich, her eyes flashing. "He could have drowned all four of them! If he doesn't obey our instructions, and the orders of the boatmen, we'll handcuff him and tie him to the raft! I don't care who he is!"

Dietrich said: "I'm sorry, madam. I'm terribly sorry. I apologize. I will speak to him."

"And," she added, her voice still seething with rage, "if he has bodyguards, order them to stay with him or we'll tie *them* to the raft, too. Do I make myself clear?"

"Yes, sir. I mean, yes ma'am." Dietrich reddened and left.

(Available in Fall 2000)

THE DINKUM DEATHS:
Murder on the Great Barrier Reef

(Barry Ross has come to Cairns, Australia, to solve two murders on Dinkum Island, part of the Great Barrier Reef. One of the scientists who worked with the dead men has disappeared, and when Joe and Julie Muck arrive, they are dispatched to find him. The trail leads into the Strzelecki Desert in Sturt National Park, in central Australia. They must find a shabby character named Geoff Billabong (obviously a fake name) working on the dingo fence, part of a system to confine the movements of wild dogs called dingos. Joe and Julie stop at Cameron Corners, where Joe stands in South Australia and drives a golf ball across New South Wales into Queensland. From there Joe and Julie head east.)

Chapter Eighteen

Joe drove their pickup along the dirt road that parallels the dingo fence. Sturt National Park stretched in all directions, its red sand dunes and mulga forests shimmered as though being steeped in boiling water.

"Must be 48 today," Julie commented, wiping her brow.

"That's 110 degrees," Joe said. "Put another ice cube down my back. I don't see how those dragon lizards take it."

"They perch in trees," Julie responded.

"That wouldn't help me," he said.

"You're not a reptile,"she replied.

The dust didn't help. Their perspiration turned brown. Lines formed on their faces and down their necks.

Ten minutes later, their search for Geoff Billabong ended. He stood beside a post, pliers gripping two strands of wire, pulling and tightening. As they drove up and stopped, he turned and mopped his brow.

They got out, expecting a friendly "Hoi myte," but got nothing more than an indifferent scowl. The worker fit the description all right. Sandy curled hair, always dirty, blue shorts turning mud color, old sandals chipped and scuffed, and a ripped tee shirt showing 4XXXX Beer. To block out the scorching sun, he wore a ragged fiber hat.

As they approached, he turned back to his work.

"Geoff?" Joe asked.

The man said nothing.

"We're from Amurrika," Joe offered, giving it an Australian pronunciation. "G-day, myte."

No response.

Julie asked: "Is this the dingo fence everybody's been telling us about?"

No response.

With this, Joe pulled out his chief weapon. "We got a couple ice cold Four Xes in the truck."

The man's hands stopped moving. His head slowly turned around as he laid eyes on them for the first time.

"Thet so?" he asked.

"Yeh," said Joe, lapsing into the lingo, "Motty keld 'n tysty nthis yer desert 'eat. We got some questions we'd loik t'esk ye."

Billabong pulled a foul blue rag out of his pocket and wiped his brow, saying: "Ye godda deal."

Grabbing the cooler they repaired to the shade of a sand hill spider flower tree. Passing around the beers, Joe started out: "We's workin' on the reef off Cairns and lookin' fer a myte."

Billabong took a big slug of beer, part of which streamed down his chin. "Yeh?"

"You used to work fer 'im. A biologist. Did some ree-search on the corals of Dinkum Island. Ring a bell?"

Geoff said nothing, but the drop of his head answered "Yes."

"'Bout two, three years ago?"

Nodding of the head.

Julie asked very simply: "What was his name?"

Billabong took another slug and asked: "He in trouble?"

Joe answered. "Nope. We just want t'esk him some things 'bout his work."

"You cops?"

"Naw. Just here on holidye. 'Specting to go to the beaches at Sydney when we get throo hyer."

"You ain't cops?"

Julie answered, lying. "We take pictures, sell 'em in magazines."

Geoff scowled again. "You ain't publishin' nuttin' on me?"

Joe shook his head vigorously. "Naw, mon. We looking fer this other chap to find out 'bout the Great Barrier Reef. That's all. No big deal."

Geoff: "He ain't hyer no more."

"We'd like to go talk with 'im 'bout the fishes and such. Know where he is?"

"Cottlesloe."

"Where's that?"

"On the beach."

"Where?"

"Perth."

Joe snorted. "Aw, shit, mon. We got t'keep on goin' t'other side o' this hyer continent?"

Julie persisted. "What's his name?"

Geoff looked at the cooler. She brought out another beer and opened it. He took a long swig, then looked at the cooler again.

She brought out a sandwich. He took it.

Then he spoke in a low tone. "Peter Murray."

With that, he took his beer and sandwich, went back to the dingo fence, set them down, and picked up his pliers again.

Joe looked at Julie. They got in the truck and drove away.

A flight of corellas shrieked as the birds flapped overhead, their white wings glistening in the desert sun. Beneath a Grevillea tree a low-slung shingle-back lizard wiggled slowly away in the shadows.

(Available in fall 2000)

Colophon

The text of *Shorty and the Radio Men* is set in 11.5 point Galliard on 13 point leading. Chapter titles are Gadget.

The text stock is 60 lb recycled paper. It is acid-free for archival durability, and it meets or exceeds all guidelines set forth by the U. S. Environmental Protection Agency for recycled content and use for post-consumer waste. The binding is Perfect Bind.

The cover, text and special character design and production are by Dan Schiffer, Digimedia, Vancouver, WA 98683, 360-885-3701. All drawings are by the author, or revised from his original works.

The book was produced on an Apple Power Macintosh 8500 using PageMaker 6.5.